SCIENCE, EVOLUTION AND CREATIONISM

HUMAN EVOLUTION

SCIENCE, EVOLUTION AND CREATIONISM

Additional books in this series can be found on Nova's website under the Series tab.

Additional e-books in this series can be found on Nova's website under the e-book tab.

SCIENCE, EVOLUTION AND CREATIONISM

HUMAN EVOLUTION

PRANVEER SINGH

nova publishers
New York

Library of Congress Cataloging-in-Publication Data

Human evolution / editor, Pranveer Singh (Assistant Professor, Department of Zoology, Indira Gandhi National Tribal University (IGNTU), Amarkantak, India).
pages cm
Includes index.
ISBN 978-1-63483-423-0 (hardcover)
1. Human evolution. I. Singh, Pranveer.
GN281.H84745 2014
599.93'8--dc23

2015026399

Published by Nova Science Publishers, Inc. † New York

Dedicated to the fond and loving memory of Aazi, Papa and Mumy

CONTENTS

PREFACE

"Any ape can reach for a banana, but only humans can reach for the stars"

VS Ramachandran

This explains the validity of the famous quote from Theodosius Dobzhansky that *"Nothing in biology makes sense except in light of evolution."* The discretionary powers of Evolution, being the supreme power, reigning the Mother Earth fascinates and enthralls because of its ability to limit the evolutionary development of creatures required by the job assigned to them. That is why, an ape or for that matter or a protozoa are not allowed to have abstract or scientific thinking thus limiting the future conflict that might have arisen to ascertain the supremacy of the living being. And this explains why a book on human evolution? What is it and how does it exclude the human from other living beings on Earth?

Donald Johanson has beautifully summarized the course of human evolution to date and its future course:

> "Where we are going as a species is a big question. Human evolution certainly hasn't stopped. Every time individuals produce a new zygote, there's a reshuffling and recombination of genes. And we don't know where all of that is going to take us".

This book on 'Human Evolution' will be a comprehensive book covering all uncovered issues relevant to the human evolution. This book deals with the 'Human Evolution' in its most exhaustive detail. Literature/work up to the latest 2013-14 have been incorporated thus including every possible information relevant to the subject.

The book is divided into different chapters highlighting morphological, anatomical, behavioral, cognitive and socio-cultural aspects in detail that has led to the evolution of primates to human. Historical perspectives and events since Darwin are discussed in detail. Various evidences, like fossil, molecular, biochemical, serological, morphological and anatomical have been taken into account to trace the evolutionary path leading to human. Theories of mankind's place of origin, its migration and subsequent colonization to different parts of the Earth are highlighted in detail. Socio-cultural, ethological and cognitive aspects that gradually transformed primates to human are discussed in detail.

This book is supplemented with a table summarizing evolution from primate to human via different intermediate species along with highlighting features of the same, and also the glossary of human evolution with important bullet points. Wherever necessary, text is supplemented with figures, ray diagrams and appendices.

This book is written in a lucid, but comprehensive manner and will make for an interesting read to the people from scientific and non-scientific backgrounds. This book will cover diverse disciplines like science (evolution), primatology (primate biology), social science, behavioral science (ethology) and anthropology for undergraduates, postgraduates and researchers alike.

To speak gratitude is courteous and pleasant, to enact gratitude is generous and noble, but to live gratitude is to touch Heaven. My most sincere gratitude is reserved for the Nova Publishers, its president and people alike for bestowing me with the repeated opportunity of writing and having faith in the quality of my work while also being flexible and accommodating to my demands. I am indebted to my family and colleagues for whatever help they had extended that had direct or indirect bearing in bringing this work to publication.

July 9, 2015

Dr. Pranveer Singh
Assistant Professor
Department of Zoology
Indira Gandhi National Tribal University (IGNTU)
Amarkantak, Madhya Pradesh (MP), India
E-mail: mailpranveer@gmail.com

Chapter 1

Introduction to Human Evolution

Abstract

The incredible story of our evolution from ape ancestors spans 6 million years or more, and features the acquirement of traits from bipedal walking, large brains, hairlessness, tool-making, hunting and harnessing fire, to the more recent development of language, art, culture and civilization.

Over the last century, many spectacular discoveries have shed light on the history of the human family. Somewhere between 12 and 19 different species of early humans are recognized, though palaeo-anthropologists bitterly dispute how they are related. Famous fossils include the remarkably complete "Lucy", dug up in Ethiopia in 1974, and the astonishing "hobbit" species, *Homo floresiensis*, found on an Indonesian island in 2004.

Introduction

Darwin's *The Origin of Species,* published in 1859, suggested that humans were descended from African apes. However, no fossils of our ancestors were discovered in Africa until 1924, when Raymond Dart dug up the "Taung child" - a 3-million to 4 million-year-old Australopithecine [1, 2].

1.1. Overview

Human evolution is the evolutionary process leading up to the appearance of modern humans from now-extinct primates. While it began with the last common ancestor of all life, the topic usually covers only the evolutionary history of primates, in particular the genus *Homo*, and the emergence of *Homo sapiens* as a distinct species of hominids (or "great apes"). The study of human evolution involves many scientific disciplines, including physical anthropology, primatology, archaeology, ethology, linguistics, evolutionary psychology, embryology and genetics [3].

Genetic studies show that primates diverged from other mammals about 85 million years ago in the Late Cretaceous period, and the earliest fossils appear in the Paleocene, around 55 million years ago. The family Hominidae diverged from the Hylobatidae (Gibbon) family 15-20 million years ago, and around 14 million years ago, the Ponginae (orangutans) diverged from the Hominidae family [4]. Bipedalism is the basic adaption of the Hominin line, and the earliest bipedal Hominin is considered to be either *Sahelanthropus* or *Orrorin*, with *Ardipithecus*, a full bipedal, coming somewhat later. The gorilla and chimpanzee diverged around the same time, about 4-6 million years ago, and either *Sahelanthropus* or *Orrorin* may be our last shared ancestor with them. The early bipedals eventually evolved into the australopithecines and later the genus *Homo*.

The earliest documented members of the genus *Homo* are *Homo habilis* which evolved around 2.3 million years ago; the earliest species for which there is positive evidence of use of stone tools. The brains of these early hominins were about the same size as that of a chimpanzee, although it has been suggested that this was the time in which the human SRGAP2 gene doubled, producing a more rapid wiring of the frontal cortex. During the next million years a process of encephalization began, and with the arrival of *Homo erectus* in the fossil record, cranial capacity had doubled to 850 cm^3. *Homo erectus* and *Homo ergaster* were the first of the hominina to leave Africa, and these species spread through Africa, Asia, and Europe between 1.3 to 1.8 million years ago. This increase in human brain size is equivalent to every generation having an additional 125,000 neurons more than their parents. It is believed that these species were the first to use fire and complex tools. According to the Recent African Ancestry theory, modern humans evolved in Africa possibly from *Homo heidelbergensis*, *Homo rhodesiensis* or *Homo antecessor* and migrated out of the continent some 50,000 to 100,000 years ago, replacing local populations of *Homo erectus*, *Homo denisova*, *Homo floresiensis* and *Homo neanderthalensis* [5, 6].

Archaic *Homo sapiens*, the forerunner of anatomically modern humans, evolved between 400,000 and 250,000 years ago (10, O'Neil). Recent DNA evidence suggests that several haplotypes of Neanderthal origin are present among all non-African populations, and Neanderthals and other hominids, such as *Denisova hominin* may have contributed up to 6% of their genome to present-day humans, suggestive of a limited inter-breeding between these species [7-9].

Anatomically modern humans evolved from archaic *Homo sapiens*in the Middle Paleolithic, about 200,000 years ago. The transition to behavioral modernity with the development of symbolic culture, language, and specialized lithic technology happened around 50,000 years ago according to many anthropologists [10] although some suggest a gradual change in behavior over a longer time span [11].

Viewed zoologically, we humans are *Homo sapiens*, a culture-bearing, upright-walking species that lives on the ground and first evolved in Africa between 100,000 and 200,000 years ago. We are now the only living members of what many zoologists refer to as the human tribe, Hominini, but there is abundant fossil evidence to indicate that we were preceded for millions of years by other hominins, such as *Australopithecus*, and that our species also lived for a time contemporaneously with at least one other member of our genus, *Homo neanderthalensis* (the Neanderthals). In addition, we and our predecessors have always shared the Earth with other apelike primates, from the modern-day gorilla to the long-extinct *Dryopithecus* [9]. That we and the extinct hominins are somehow related and that we and the apes, both living and extinct, are also somehow related is accepted by anthropologists and biologists everywhere. Yet the exact nature of our evolutionary relationships has been the subject of debate and investigation since the great British naturalist Charles Darwin published his monumental books *On the Origin of Species* (1859) and *The Descent of Man* (1871) [1]. Darwin never claimed, as some of his Victorian contemporaries insisted he had, that "man was descended from the apes," and modern scientists would view such a statement as a useless simplification—just as they would dismiss any popular notions that a certain extinct species is the "missing link" between man and the apes. There is theoretically, however, a common ancestor that existed millions of years ago.

This ancestral species does not constitute a "missing link" along a lineage but rather a node for divergence into separate lineages. This ancient primate has not been identified and may never be known with certainty, because fossil relationships are unclear even within the human lineage, which is more recent. In fact, the human "family tree" may be better described as a "family bush," within which it is impossible to connect a full chronological series of species, leading to *Homo sapiens* that experts can agree upon [1].

The primary resource for detailing the path of human evolution will always be fossil specimens. Certainly, the trove of fossils from Africa and Eurasia indicates that, unlike today, more than one species of our family has lived at the same time for most of human history. The nature of specific fossil specimens and species can be accurately described, as can the location where they were found and the period of time when they lived (Figure 1); but questions of how species lived and why they might have either died out or evolved into other species can only be addressed by formulating scenarios, albeit scientifically informed ones [11].

These scenarios are based on contextual information gleaned from localities where the fossils were collected. In devising such scenarios and filling in the human family bush, researchers must consult a large and diverse array of fossils, and they must also employ refined excavation methods and records, geochemical dating techniques, and data from other specialized fields such as genetics, ecology and paleoecology, and ethology (animal behaviour)—in short, all the tools of the multidisciplinary science of paleoanthropology.

Figure 1. Fossil Hominid Evolution Display at the Museum of Osteology, Oklahoma City, Oklahoma, USA.

1.2. Chronology of Study

This chapter is a discussion of the broad career of the human tribe from its probable beginnings millions of years ago in the Miocene Epoch to the development of tool-based and symbolically structured modern human culture only tens of thousands of years ago, during the geologically recent Pleistocene Epoch.

Particular attention is paid to the fossil evidence for this history and to the principal models of evolution that have gained the most credence in the scientific community and evolution for a full explanation of evolutionary theory, including its main proponents both before and after Darwin, its arousal of both resistance and acceptance in society, and the scientific tools used to investigate the theory and prove its validity [10].

1.3. Pre-Darwin Era

The word *homo*, the name of the biological genus to which humans belong, is Latin for "human." It was chosen originally by Carolus Linnaeus in his classification system. The word "human" is from the Latin *humanus*, the adjectival form of *homo*. The Latin "homo" derives from the Indo-European root **dhghem*, or "earth." Linnaeus and other scientists of his time also considered the great apes to be the closest relatives of humans based on morphological and anatomical similarities [1].

1.4. Darwinian Era

The possibility of linking humans with earlier apes by descent became clear only after 1859 with the publication of Charles Darwin's *On the Origin of Species*, in which he argued for the idea of the evolution of new species from earlier ones. Darwin's book did not address the question of human evolution, saying only that "Light will be thrown on the origin of man and his history."

The first debates about the nature of human evolution arose between Thomas Huxley and Richard Owen. Huxley argued for human evolution from apes by illustrating many of the similarities and differences between humans and apes, and did so particularly in his 1863 book *Evidence as to Man's Place in Nature*. However, many of Darwin's early supporters (such as Alfred Russel Wallace and Charles Lyell) did not initially agree that the origin of the mental capacities and the moral sensibilities of humans could be explained by natural selection, though this later changed. Darwin applied the theory of evolution and sexual selection to humans when he published *The Descent of Man* in 1871 [1].

1.5. Earliest Fossils

A major problem at that time was the lack of fossil intermediaries. Despite the 1891 discovery by Eugène Dubois of what is now called *Homo erectus* at Trinil, Java, it was only in the 1920s when such fossils were discovered in Africa, that intermediate species began to accumulate. In 1925, Raymond Dart described *Australopithecus africanus*. The type specimen was the Taung Child, an Australopithecine infant which was discovered in a cave. The child's remains were a remarkably well-preserved tiny skull and an endocranial cast of the brain (Figure 2).

Figure 2. Louis Leakey examining skulls from Olduvai Gorge, Tanzania.

Although the brain was small (410 cm^3), its shape was rounded, unlike that of chimpanzees and gorillas, and more like a modern human brain. Also, the specimen showed short canine teeth, and the position of the foramen magnum was evidence of bipedal locomotion. All of these traits convinced Dart that the Taung baby was a bipedal human ancestor, a transitional form between apes and humans (Figure 2).

1.6. East Africa, Hot-Bed of Human Evolution

During the 1960s and 1970s, hundreds of fossils were found, particularly in East Africa in the regions of the Olduvai gorge and Lake Turkana. The driving force in the East African researches was the Leakey family, with Louis Leakey and his wife Mary Leakey, and later their son Richard and daughter in-law Meave being among the most successful fossil hunters and palaeoanthropologists. From the fossil beds of Olduvai and Lake Turkana they amassed fossils of australopithecines, early *Homo* and even *Homo erectus*.

These finds cemented Africa as the cradle of humankind. In the 1980s, Ethiopia emerged as the new hot spot of palaeoanthropology as "Lucy", the most complete fossil member of the species *Australopithecus afarensis*, was found by Donald Johanson in Hadar in the desertic Middle Awash region of northern Ethiopia. This area would be the location of many new hominin fossils, particularly those uncovered by the teams of Tim White in the 1990s, such as *Ardipithecus ramidus* (Figure 2).

1.7. The Genetic Revolution

The genetic revolution in studies of human evolution started when Vincent Sarich and Allan Wilson measured the strength of immunological cross-reactions of blood serum albumin between pairs of creatures, including humans and African apes (chimpanzees and gorillas) [12]. The strength of the reaction could be expressed numerically as an immunological distance, which was in turn proportional to the number of amino acid differences between homologous proteins in different species. By constructing a calibration curve of the ID of species' pairs with known divergence times in the fossil record, the data could be used as a molecular clock to estimate the times of divergence of pairs with poorer or unknown fossil records.

In their seminal 1967 paper in *Science*, Sarich and Wilson estimated the divergence time of humans and apes as four to five million years ago [12], at a time when standard interpretations of the fossil record gave this divergence as at least 10 to as much as 30 million years. Subsequent fossil discoveries, notably Lucy, and reinterpretation of older fossil materials, notably *Ramapithecus*, showed the younger estimates to be correct and validated the albumin method. Application of the molecular clock principle revolutionized the study of molecular evolution.

1.8. PRIMATE EVOLUTION

Evolutionary history of the primates can be traced back 65 million years. The oldest known primate-like mammal species, the *Plesiadapis*, came from North America, but they were widespread in Eurasia and Africa during the tropical conditions of the Paleocene and Eocene.

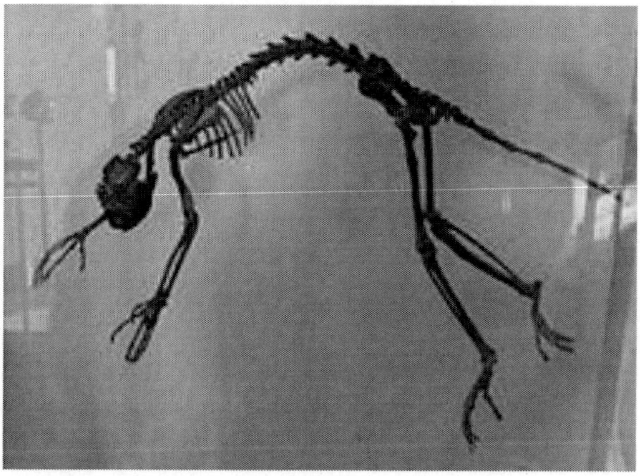

Figure 3. *Notharctus.*

David Begun concluded that early primates flourished in Eurasia and that a lineage leading to the African apes and humans, including *Dryopithecus*, migrated south from Europe or Western Asia into Africa. The surviving tropical population of primates, which is seen most completely in the upper Eocene and lowermost Oligocene fossil beds of the Faiyum depression southwest of Cairo, gave rise to all living species—lemurs of Madagascar, lorises of Southeast Asia, galagos or "bush babies" of Africa, and the anthropoids: platyrrhine or New World monkeys, catarrhines or Old World monkeys, and the great apes, including humans [13].

The earliest known catarrhine is *Kamoyapithecus* from uppermost Oligocene at Eragaleit in the northern Kenya Rift Valley, dated to 24 million years ago [14]. Its ancestry is thought to be species related to *Aegyptopithecus*, *Propliopithecus*, and *Parapithecus* from the Fayum, at around 35 million years ago [15]. In 2010, *Saadanius* was described as a close relative of the last common ancestor of the crown catarrhines, and tentatively dated to 29–28 million years ago, helping to fill an 11-million-year gap in the fossil record [16].

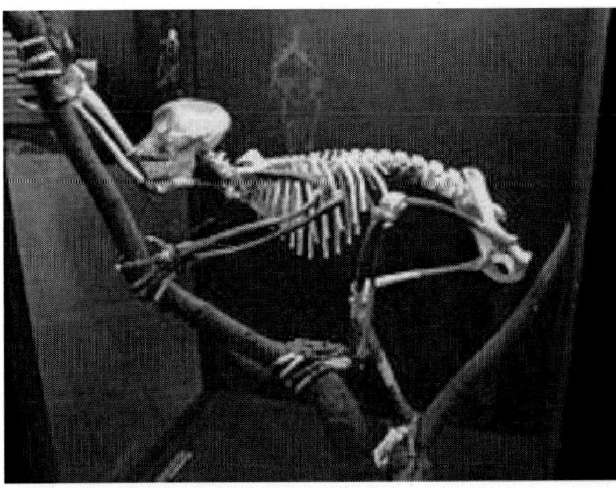

Figure 4. Reconstructed tailless *Proconsul* skeleton.

In the early Miocene, about 22 million years ago, the many kinds of arboreally adapted primitive catarrhines from East Africa suggest a long history of prior diversification. Fossils at 20 million years ago include fragments attributed to *Victoriapithecus*, the earliest Old World Monkey. Among the genera thought to be in the ape lineage leading up to 13 million years ago are *Proconsul*, *Rangwapithecus*, *Dendropithecus*, *Limnopithecus*, *Nacholapithecus*, *Equatorius*, *Nyanzapithecus*, *Afropithecus*, *Heliopithecus*, and *Kenyapithecus*, all from East Africa (Figures 3, 4).

The presence of other generalized non-cercopithecids of middle Miocene age from sites far distant—*Otavipithecus* from cave deposits in Namibia, and *Pierolapithecus* and *Dryopithecus* from France, Spain and Austria—is evidence of a wide diversity of forms across Africa and the Mediterranean basin during the relatively warm and equable climatic regimes of the early and middle Miocene. The youngest of the Miocene hominoids, *Oreopithecus*, is from coal beds in Italy that have been dated to 9 million years ago.

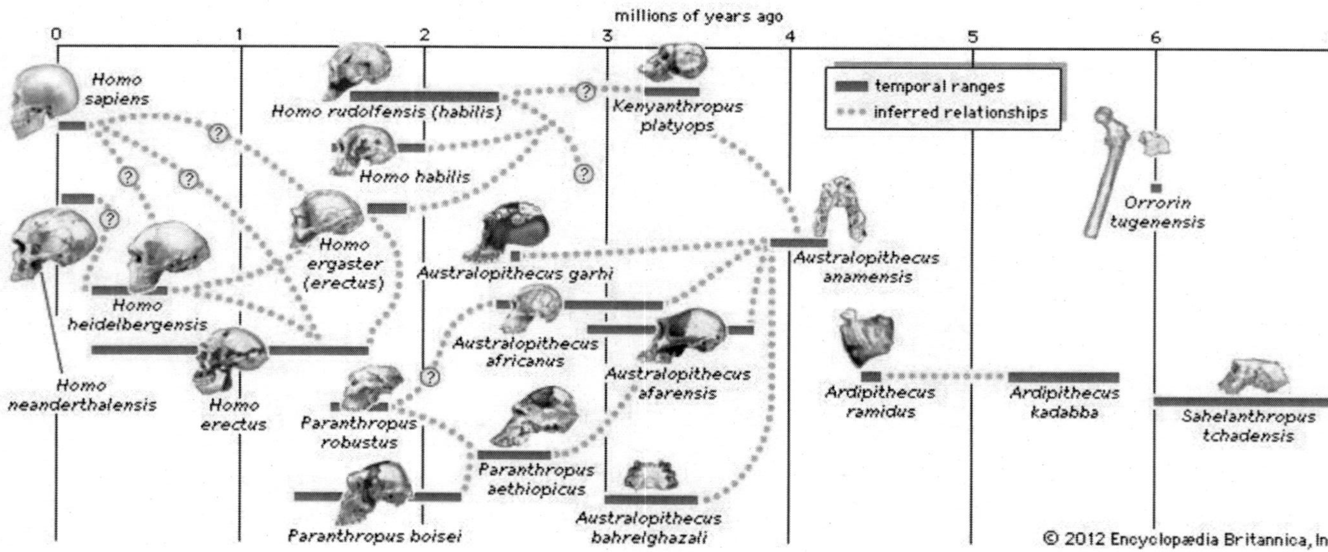

Figure 5. Possible pathways in the evolution of human lineages (Adapted from Human evolution. 2015. *Encyclopedia Britannica Online*. Retrieved 18 June, 2015, from http://www.britannica.com/science/human-evolution) [17].

Molecular evidence indicates that the lineage of gibbons (family Hylobatidae) diverged from Great Apes some 18-12 million years ago, and that of orangutans (subfamily Ponginae) diverged from the other Great Apes at about 12 million years; there are no fossils that clearly document the ancestry of gibbons, which may have originated in a so-far-unknown South East Asian hominoid population, but fossil proto-orangutans may be represented by *Sivapithecus* from India and *Griphopithecus* from Turkey, dated to around 10 million years ago.

1.9. Divergence of the Human Clade from Other Great Apes

Species close to the last common ancestor of gorillas, chimpanzees and humans may be represented by *Nakalipithecus* fossils found in Kenya and *Ouranopithecus* found in Greece. Molecular evidence suggests that between 8 and 4 million years ago, first the gorillas, and then the chimpanzees (genus *Pan*) split off from the line leading to the humans; human DNA is approximately 98.4% identical to that of chimpanzees when comparing single nucleotide polymorphisms (see human evolutionary genetics). The fossil record of gorillas and chimpanzees is limited. Both poor preservation (rain forest soils tend to be acidic and dissolve bone) and sampling bias probably contribute to this problem (Figure 5) [17].

Other hominins likely adapted to the drier environments outside the equatorial belt, along with antelopes, hyenas, dogs, pigs, elephants, and horses. The equatorial belt contracted after about 8 million years ago. There is very little fossil evidence for the split of the hominin lineage from the lineages of gorillas and chimpanzees. The earliest fossils that have been argued to belong to the human lineage are *Sahelanthropus tchadensis* (7 Ma) and *Orrorin tugenensis* (6 Ma), followed by *Ardipithecus* (5.5–4.4 Ma), with species *Ar. kadabba* and *Ar. ramidus*.

References

[1] Darwin C (1871). This edition published Introduction by John Tyler Bonner & Robert M. May). *The Descent of Man, and Selection in Relation to Sex*. Princeton NJ: Princeton University Press.

[2] Dart RA (1925), *Australopithecus africanus*: The Man-Ape of South Africa, *Nature* 115, 195–99.

[3] Heng HH (2009). The genome-centric concept: resynthesis of evolutionary theory. *BioEssays* 31, 512–25.

[4] Dawkins R (2004). The Ancestor's Tale. Query: Hominidae/ Hylobatidae. *Time Tree*.

[5] Stringer CB (1994). Evolution of Early Humans. In Steve Jones, Robert Martin, David Pilbeam. *The Cambridge Encyclopedia of Human Evolution*. Cambridge: Cambridge University Press. p. 242.

[6] McHenry H M (2009). Human Evolution. In Michael Ruse & Joseph Travis. *Evolution: the first four billion years*. Cambridge, Massachusetts: The Belknap Press of Harvard University Press. p. 265.

[7] Reich D, Green RE, Kircher M, et al. (2010). Genetic history of an archaic hominin group from Denisova Cave in Siberia. *Nature* 468, 1053–60.

[8] Noonan JP (2010). Neanderthal genomics and the evolution of modern humans. Genome *Res*. 20, 547–53.

[9] Abi-Rached L, Jobin M J, Kulkarni S, McWhinnie A Dalva et al. (2011). The shaping of modern human immune systems by multiregional admixture with archaic humans. *Science* 334, 89–94.

[10] Mellars P (2006). Why did modern human populations disperse from Africa ca. 60,000 years ago?. *Proc. Natl. Acad. Sc. USA* 103, 9381–6.

[11] Mcbrearty S, Brooks AS (2000). The revolution that wasn't: a new interpretation of the origin of modern human behavior. *J. Hum. Evol.* 39, 453–563.

[12] Sarich VM, Wilson AC (1967). Immunological time scale for hominid evolution. *Science* 158, 1200–1203.

[13] Kordos L, Begun D R (2001). Primates from Rudabánya: allocation of specimens to individuals, sex and age categories. *J. Hum. Evol.* 40, 17–39.

[14] Cameron DC (2004). *Hominid adaptations and extinctions*. UNSW Press. p. 76.

[15] Wallace DR (2005). *Beasts of Eden: Walking Whales, Dawn Horses, and Other Enigmas of Mammal Evolution*. University of California Press. pp. 240.

[16] Zalmout IS, Sanders WJ, MacLatchy LM, et al. (2010). New Oligocene primate from Saudi Arabia and the divergence of apes and Old World Monkeys. *Nature* 466, 360–364.

[17] Human evolution. 2015. *Encyclopedia Britannica Online*. Retrieved 18 June, 2015, from http://www.britannica.com/science/human-evolution.

Chapter 2

EVOLUTION AND DISPERSAL OF *HOMININS*

ABSTRACT

Homo sapiens are the only extant species of its genus, *Homo*. While some other, extinct *Homo* species might have been ancestors of *Homo sapiens*, many were likely our "cousins", having speciated away from our ancestral line. Based on archaeological and paleontological evidence, it has been possible to infer, to some extent, the ancient dietary practices of various *Homo* species and to study the role of diet in physical and behavioral evolution within *Homo*.

INTRODUCTION

2.1. GENUS AUSTRALOPITHECUS

The *Australopithecus* genus evolved in eastern Africa around 4 million years ago before spreading throughout the continent and eventually becoming extinct 2 million years ago. During this time period various forms of australopiths existed, including *Australopithecus anamensis*, *A. afarensis*, *A. sediba*, and *A. africanus*. There is still some debate amongst academics whether certain African hominid species of this time, such as *A. robustus* and *A. boisei*, constitute members of the same genus; if so, they would be considered to be *robust australopiths* whilst the others would be considered *gracile australopiths*. However, if these species do indeed constitute their own genus, then they may be given their own name, the *Paranthropus*.

- Australopithecus (4–1.8 Ma), with species *Au. anamensis*, *Au. afarensis, Au. africanus, Au. bahrelghazali, Au. garhi*, and *Au. sediba* (Figure 1);
- Kenyanthropus (3–2.7 Ma), with species Kenyanthropus platyops;
- Paranthropus (3–1.2 Ma), with species *P. aethiopicus, P. boisei,* and *P. robustus*

Figure 1. A reconstruction of a female *Australopithecus afarensis* (National Museum of Natural History).

2.2. GENUS *HOMO*

Homo sapiens are the only extant species of its genus, *Homo*. While some other, extinct *Homo* species might have been ancestors of *Homo sapiens*, many were likely our "cousins", having speciated away from our ancestral line [1]. There is not yet a consensus as to which of these groups should count as separate species and which as subspecies. In some cases this is due to the dearth of fossils, in other cases it is due to the slight differences used to classify species in the *Homo* genus [2]. The Sahara pump theory (describing an occasionally passable "wet" Sahara Desert) provides one possible explanation of the early variation in the genus *Homo* (Figures 2, 4).

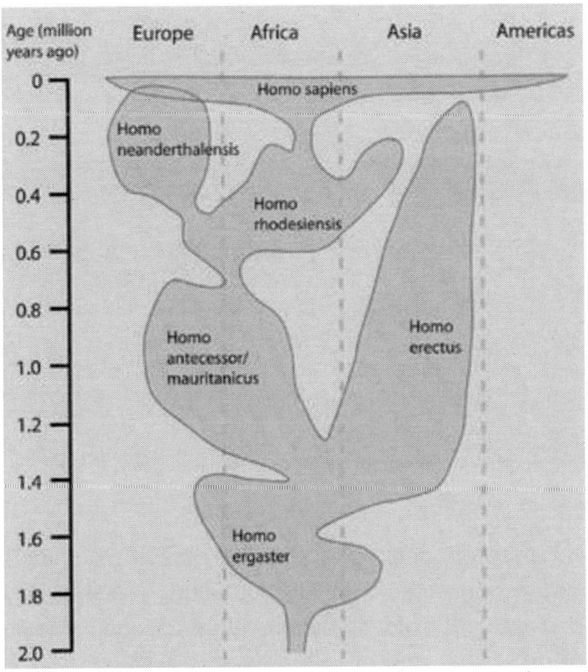

Figure 2. One current view of the temporal and geographical distribution of hominid populations (Adapted from Reed et al. 2004, *PLoS Biol* 2, e340). Other interpretations differ mainly in the taxonomy and geographical distribution of hominid species.

Figure 3. A reconstruction of *Homo habilis.*

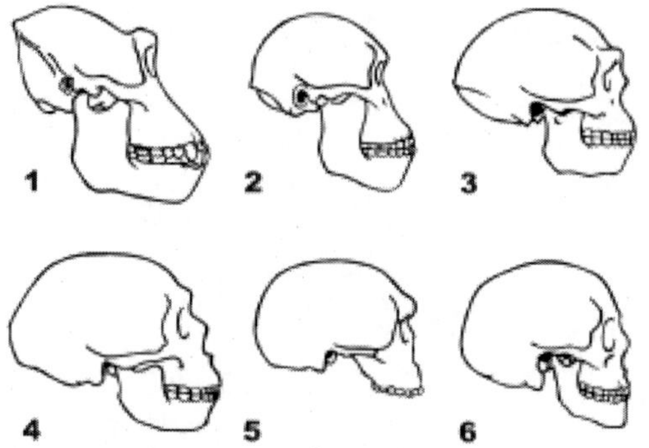

Figure 4. Skulls of 1. *Gorilla* 2. *Australopithecus* 3. *Homo erectus* 4. Neanderthal (La Chapelle aux Saints) 5. Steinheim Skull 6. modern *Homo sapiens*. Notice the decreasing prognathism and thickness of the brow ridge, and the increasing size of the forehead.

Based on archaeological and paleontological evidence, it has been possible to infer, to some extent, the ancient dietary practices of various *Homo* species and to study the role of diet in physical and behavioral evolution within *Homo* [3-6].

According to the Toba catastrophe theory to which some anthropologists and archeologists subscribe, the super eruption of Lake Toba on Sumatra island in Indonesia roughly 70,000 years ago had global consequences, killing most humans then alive and creating a population bottleneck that affected the genetic inheritance of all humans today.

2.3. HOMO *HABILIS* AND HOMO GAUTENGENSIS

Homo habilis lived from about 2.4 to 1.4 Ma. *Homo habilis* evolved in South and East Africa in the late Pliocene or early Pleistocene, 2.5–2 Ma, when it diverged from the australopithecines. *Homo habilis* had smaller molars and larger brains than the australopithecines, and made tools from stone and perhaps animal bones (Figure 4).

One of the first known hominids, it was nicknamed ‘handy man’ by discoverer Louis Leakey due to its association with stone tools. Some scientists have proposed moving this species out of *Homo* and into *Australopithecus* due to the morphology of its skeleton being more adapted to living on trees rather than to moving on two legs like *Homo sapiens* [7].

It was considered to be the first species of the genus *Homo* until May 2010, when a new species, *Homo gautengensis* was discovered in South Africa (Figure 2), that most likely arose earlier than *Homo habilis* [7, 8].

2.4. HOMO *RUDOLFENSIS* AND HOMO *GEORGICUS*

These are proposed species names for fossils from about 1.9–1.6 Ma, whose relation to *Homo habilis* is not yet clear.

Homo rudolfensis refers to a single, incomplete skull from Kenya. Scientists have suggested that this was another *Homo habilis*, but this has not been confirmed.

Homo georgicus, from Georgia, may be an intermediate form between *Homo habilis* and *Homo erectus*, or a sub-species of *Homo erectus* [9, 10].

2.5. HOMO *ERGASTER* AND HOMO *ERECTUS*

The first fossils of *Homo erectus* were discovered by Dutch physician Eugene Dubois in 1891 on the Indonesian island of Java (Figure 2). He originally named the material *Pithecanthropus erectus* based on its morphology, which he considered to be intermediate between that of humans and apes [11]. *Homo erectus* (*H. erectus*) lived from about 1.8 Ma to about 70,000 years ago (which would indicate that they were probably wiped out by the Toba catastrophe; however, *Homo erectus soloensis* and *Homo floresiensis* survived it). Often the early phase, from 1.8 to 1.25 Ma, is considered to be a separate species, *Homo ergaster*, or it is seen as a subspecies of *Homo erectus*, *Homo erectus ergaster* [12].

In the early Pleistocene, 1.5–1 Ma, in Africa some populations of *Homo habilis* are thought to have evolved larger brains and made more elaborate stone tools; these differences and others are sufficient for anthropologists to classify them as a new species, *Homo erectus* [12]. This was made possible by the evolution of locking knees and a different location of the foramen magnum (the hole in the skull where the spine enters). They may have used fire to cook their meat.

A famous example of *Homo erectus* is Peking man; others were found in Asia (notably in Indonesia), Africa, and Europe. Many paleoanthropologists now use the term *Homo ergaster* for the non-Asian forms of this group, and reserve *Homo erectus* only for those fossils that are found in Asia and meet certain skeletal and dental requirements which differ slightly from *H. ergaster*.

2.6. HOMO *CEPRANENSIS* AND HOMO *ANTECESSOR*

These are proposed as species that may be intermediate between *H. erectus* and *H. heidelbergensis*.

H. antecessor is known from fossils from Spain and England that are dated 1.2 Ma–500 ka [13, 14].

H. cepranensis refers to a single skull cap from Italy, estimated to be about 800,000 years old [15].

2.7. HOMO HEIDELBERGENSIS

Figure 5. Reconstruction of *Homo heidelbergensis* which may be the direct ancestor of both *Homo neanderthalensis* and *Homo sapiens*.

H. heidelbergensis (Heidelberg Man) lived from about 800,000 to about 300,000 years ago (Figures 2, 5). Also proposed as *Homo sapiens heidelbergensis* or *Homo sapiens paleohungaricus* [16].

2.8. *Homo rhodesiensis*, and the Gawis Man

H. rhodesiensis, estimated to be 300,000–125,000 years old. Most current researchers place Rhodesian man within the group of *Homo heidelbergensis*, though other designations such as Archaic *Homo sapiens* and *Homo sapiens rhodesiensis* have been proposed.

In February 2006 a fossil, the Gawis cranium, was found which might possibly be a species intermediate between *H. erectus* and *H. sapiens* or one of many evolutionary dead ends. The skull from Gawis, Ethiopia, is believed to be 500,000–250,000 years old. Only summary details are known, and the finders have not yet released a peer-reviewed study. Gawis man's facial features suggest its being either an intermediate species or an example of a "Bodo man" female.

2.9. Neanderthal and Denisovan

Figure 6. Dermoplastic reconstruction of a Neanderthal.

H. neanderthalensis, alternatively designated as *Homo sapiens neanderthalensis* [17], lived in Europe and Asia from 400,000 [18] to about 30,000 years ago (Figures 2, 6). Evidence from sequencing mitochondrial DNA indicated that no significant gene flow occurred between *H. neanderthalensis* and *H. sapiens*, and, therefore, the two were separate species that shared a common ancestor about 660,000 years ago [19-22]. However, the 2010 sequencing of the Neanderthal genome indicated that Neanderthals did indeed interbreed with anatomically modern humans *circa* 45,000 to 80,000 years ago (at the approximate time that modern humans migrated out from Africa, but before they dispersed into Europe, Asia and elsewhere) [23].

Nearly all modern non-African humans have 1% to 4% of their DNA derived from Neanderthal DNA, and this finding is consistent with recent studies indicating that the divergence of some human alleles dates to one Ma, although the interpretation of these studies has been questioned [24, 25]. Neanderthals and *Homo sapiens* could have co-existed in Europe for as long as 10,000 years, during which human populations exploded vastly outnumbering Neanderthals, possibly outcompeting them by sheer numerical strength [26].

In 2008, archaeologists working at the site of Denisova Cave in the Altai Mountains of Siberia uncovered a small bone fragment from the fifth finger of a juvenile member of Denisovans [27]. Artifacts, including a bracelet, excavated in the cave at the same level were carbon dated to around 40,000 BP. As DNA had survived in the fossil fragment due to the cool climate of the Denisova Cave, both mtDNA and nuclear genomic DNA were sequenced [28, 29].

While the divergence point of the mtDNA was unexpectedly deep in time [30], the full genomic sequence suggested the Denisovans belonged to the same lineage as Neanderthals, with the two diverging shortly after their line split from that lineage giving rise to modern humans [29].

Modern humans are known to have overlapped with Neanderthals in Europe for more than 10,000 years, and the discovery raises the possibility that Neanderthals, modern humans and the Denisova hominin may have co-existed. The existence of this distant branch creates a much more complex picture of humankind during the Late Pleistocene than previously thought [28, 31]. Evidence has also been found that as much as 6% of the genomes of some modern Melanesians derive from Denisovans, indicating limited interbreeding in Southeast Asia [32, 33].

Alleles thought to have originated in Neanderthal and the Denisova hominin have been identified at several genetic loci in the genomes of modern humans outside of Africa. HLA types from Denisovans and Neanderthal represent more than half the HLA alleles of modern Eurasians [34], indicating strong positive selection for these introgressed alleles.

2.10. HOMO FLORESIENSIS

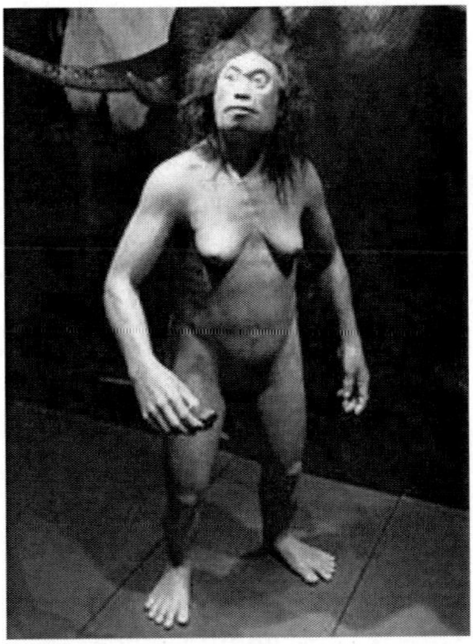

Figure 7. Restoration model of Homo floresiensis.

H. floresiensis, which lived from approximately 100,000 to 12,000 before present, has been nicknamed *hobbit* for its small size, possibly a result of insular dwarfism as shown in Figure 7 [35]. *H. floresiensis* is intriguing both for its size and its age, being an example of a recent species of the genus *Homo* that exhibits derived traits not shared with modern humans. In other words, *H. floresiensis* shares a common ancestor with modern humans, but split from the modern human lineage and followed a distinct evolutionary path. The main find was a skeleton believed to be a woman of about 30 years of age. Found in 2003 it has been dated to approximately 18,000 years old. The living woman was estimated to be one meter in height, with a brain volume of just 380 cm3 (considered small for a chimpanzee and less than a third of the *H. sapiens* average of 1400 cm3).

However, there is an ongoing debate over whether *H. floresiensis* is indeed a separate species [36]. Some scientists hold that *H. floresiensis* was a modern *H. sapiens* with pathological dwarfism [37]. This hypothesis is supported in part, because some modern humans who live on Flores, the island where the skeleton was found, are pygmies. This, coupled with pathological dwarfism, could possibly create a hobbit-like human. The other major attack on *H. floresiensis* is that it was found with tools only associated with *H. sapiens*.

The hypothesis of pathological dwarfism, however, fails to explain additional anatomical features that are unlike those of modern humans (diseased or not) but much like those of ancient members of our genus. Aside from cranial features, these features include the form of bones in the wrist, forearm, shoulder, knees, and feet. Additionally, this hypothesis fails to explain the find of multiple examples of individuals with these same characteristics, indicating they were common to a large population, and not limited to one individual.

2.11. *HOMO SAPIENS*

H. sapiens (the adjective *sapiens* is Latin for "wise" or "intelligent") have lived from about 250,000 years ago to the present. Between 400,000 years ago and the second interglacial period in the Middle Pleistocene, around 250,000 years ago (Figure 2), the trend in skull expansion and the elaboration of stone tool technologies developed, providing evidence for a transition from *H. erectus* to *H. sapiens*. The direct evidence suggests there were a migration of *H. erectus* out of Africa, then a further speciation of *H. sapiens* from *H. erectus* in Africa. A subsequent migration within and out of Africa eventually replaced the earlier dispersed *H. erectus*. This migration and origin theory is usually referred to as the *recent single origin* or Out of Africa theory. Current evidence does not preclude some multiregional evolution or some admixture of the migrant *H. sapiens* with existing *Homo* populations. This is a hotly debated area of paleo-anthropology.

Current research has established that humans are genetically highly homogenous; that is, the DNA of individuals is more alike than usual for most species, which may have resulted from their relatively recent evolution or the possibility of a population bottleneck resulting from cataclysmic natural events such as the Toba catastrophe [38, 39]. Distinctive genetic characteristics have arisen, however, primarily as the result of small groups of people moving into new environmental circumstances. These adapted traits are a very small component of the *Homo sapiens* genome, but include various characteristics such as skin color and nose form, in addition to internal characteristics such as the ability to breathe more efficiently at high altitudes.

H. sapiens idaltu, from Ethiopia, is an extinct sub-species from about 160,000 years ago.

2.12. The Quest for the *Earliest Hominin*

In the 1990s, several teams of paleoanthropologists were working throughout Africa looking for evidence of the earliest divergence of the Hominin lineage from the great apes. In 1994, Meave Leakey discovered *Australopithecus anamensis*. The find was overshadowed by Tim White's 1995 discovery of *Ardipithecus ramidus*, which pushed back the fossil record to 4.2 million years ago [40].

In 2000, Martin Pickford and Brigitte Senut discovered in the Tugen Hills of Kenya a 6-million-year-old bipedal hominin which they named *Orrorin tugenensis*. And in 2001, a team led by Michel Brunet discovered the skull of *Sahelanthropus tchadensis* which was dated as 7.2 million years ago, and which Brunet argued was a bipedal, and therefore a hominid.

2.13. Human Dispersal

Anthropologists in the 1980s were divided regarding some details of reproductive barriers and migratory dispersals of the *Homo* genus. Subsequently, genetics has been used to investigate and resolve these issues. According to the Sahara pump theory evidence suggests that genus *Homo* have migrated out of Africa at least three times (e.g., *Homo erectus*, *Homo heidelbergensis* and *Homo sapiens*).

2.14. The Out-of-Africa Model

This model proposed that modern *H. sapiens* speciated in Africa recently (approx. 200,000 years ago) and the subsequent migration through Eurasia resulted in nearly complete replacement of other *Homo* species. This model has been developed by Chris Stringer and Peter Andrews [41]. In contrast, the multiregional hypothesis proposed that *Homo* genus contained only a single interconnected population as it does today (not separate species), and that its evolution took place worldwide continuously over the last couple million years. This model was proposed in 1988 by Milford H. Wolpoff [42, 43].

Progress in DNA sequencing, specifically mitochondrial DNA (mtDNA) and then Y-chromosome DNA advanced the understanding of human origins [44-46]. Sequencing mtDNA and Y-DNA sampled from a wide range of indigenous populations revealed ancestral information relating to both male and female genetic heritage [47]. Aligned in genetic tree differences were interpreted as supportive of a recent single origin [48]. Analyses have shown a greater diversity of DNA patterns throughout Africa, consistent with the idea that Africa is the ancestral home of mitochondrial Eve and Y-chromosomal Adam [49].

Out of Africa has gained support from research using female mitochondrial DNA (mtDNA) and the male Y chromosome. After analysing genealogy trees constructed using 133 types of mtDNA, researchers concluded that all were descended from a female African progenitor, dubbed Mitochondrial Eve. Out of Africa is also supported by the fact that mitochondrial genetic diversity is highest among African populations [50].

A broad study of African genetic diversity, headed by Sarah Tishkoff, found the San people had the greatest genetic diversity among the 113 distinct populations sampled, making them one of 14 "ancestral population clusters."

The research also located the origin of modern human migration in south-western Africa, near the coastal border of Namibia and Angola [51]. The fossil evidence was insufficient for Richard Leakey to resolve this debate [40]. Studies of haplogroups in Y-chromosomal DNA and mitochondrial DNA have largely supported a recent African origin [52]. Evidence from autosomal DNA also predominantly supports a Recent African origin. However evidence for archaic admixture in modern humans had been suggested by some studies [53].

Recent sequencing of Neanderthal [21] and Denisovan [29] genomes shows that some admixture occurred. Modern humans outside Africa have 2-4% Neanderthal alleles in their genome, and some Melanesians have an additional 4-6% of Denisovan alleles.

These new results do not contradict the Out of Africa model, except in its strictest interpretation. After recovery from a genetic bottleneck that might be due to the Toba supervolcano catastrophe, a fairly small group left Africa and briefly interbred with Neanderthals, probably in the middle-east or even North Africa before their departure. Their still predominantly African descendants spread to populate the world. A fraction in turn interbred with Denisovans, probably in south-east Asia, before populating Melanesia [33]. HLA haplotypes of Neanderthal and Denisova origin have been identified in modern Eurasian and Oceanian populations [34].

2.15. Multiple Dispersal Model

There are still differing theories on whether there was a single exodus or several. A multiple dispersal model involves the Southern Dispersal theory, which has gained support in recent years from genetic, linguistic and archaeological evidence. In this theory, there was a coastal dispersal of modern humans from the Horn of Africa around 70,000 years ago. This group helped to populate Southeast Asia and Oceania, explaining the discovery of early human sites in these areas much earlier than those in the Levant.

A second wave of humans may have dispersed across the Sinai Peninsula into Asia, resulting in the bulk of human population for Eurasia. This second group possibly possessed a more sophisticated tool technology and was less dependent on coastal food sources than the original group. Much of the evidence for the first group's expansion would have been destroyed by the rising sea levels at the end of each glacial maximum. The multiple dispersal models is contradicted by studies indicating that the populations of Eurasia and the populations of Southeast Asia and Oceania are all descended from the same mitochondrial DNA lineages, which support a single migration out of Africa that gave rise to all non-African populations [54, 55].

References

[1] Strait DS, Grine FE, Moniz MA (1997). A reappraisal of early hominid phylogeny. *J. Hum. Evol.* 32, 17–82.

[2] Bryson B (2004). C 28. The Mysterious Biped. *A Short History of Nearly Everything*. Random House, Inc. pp. 522–543.

[3] Barry B (1997). The evolution of human nutrition. In: Romanucci-Ross L, Moerman DE, Tancredi LR (Eds.) *The Anthropology of Medicine: From Culture to Method*. South Hadley MA: Bergen and Garvey. pp. 96–142.

[4] Ungar P, Teaford MF (2002). *Human Diet: Its Origin and Evolution*. Westport, CT: Bergin & Garvey. p. 206.

[5] Barnicot NA (2005). Human nutrition: evolutionary perspectives. *Integr Physiol Behav Sci* 40, 114–17.

[6] Walker A (2006). Early Hominin Diets: Overview and Historical Perspectives. InPeter Ungar. *Evolution of the Human Diet: The Known, the Unknown, and the Unknowable*. US: Oxford University Press. pp. 3–10.

[7] Wood B, Collard M (1999a). The changing face of Genus Homo. *Evol. Anth.* 8(6) 195-207.

[8] Wood B (1999b). Homo rudolfensis' Alexeev, 1986-fact or phantom? *J. Hum. Evol.* 36(1): 115–8.

[9] Gabounia L, de Lumley M, Vekua A et al. (2002). Discovery of a new hominid at Dmanisi (Transcaucasia, Georgia). *Comptes Rendus Palevol* 1, 243–53.

[10] Lordkipanidze D, Vekua A, Ferring R et al. (2006). A fourth hominin skull from Dmanisi, Georgia. *The anatomical record. Part A, Discoveries in mol cell evol biol* 288, 1146–57.

[11] Turner W (1895). On M. Dubois' Description of Remains recently found in Java, named by him Pithecanthropus erectus: With Remarks on so-called Transitional Forms between Apes and Man. *J Anat Phys* 29, 424–45.

[12] Spoor F, Wood B, Zonneveld F, Zonneveld W (1994). Implications of early hominid labyrinthine morphology for evolution of human bipedal locomotion. *Nature* 369, 645–8.

[13] Bermúdez de Castro JM et al. (1997). A Hominid from the Lower Pleistocene of Atapuerca, Spain: Possible Ancestor to Neandertals and Modern Humans. Science 276, 1392-1395.

[14] Carbonell E, José M Bermúdez de Castro et al. (2008). The first hominin of Europe. *Nature* 452, 465–469.

[15] Manzi G, Mallegni F, Ascenzi A, Mallegni A (2001). A cranium for the earliest Europeans: Phylogenetic position of the hominid from Ceprano, Italy. *Proc. Natl. Acad. Sci. U.S.A.* 98, 10011–6.

[16] Czarnetzki A, Jakob T, Pusch CM (2003). Palaeopathological and variant conditions of the Homo heidelbergensis type specimen (Mauer, Germany). *J. Hum. Evol.* 44, 479–95.

[17] Harvati K (2003). The Neanderthal taxonomic position: models of intra- and inter-specific craniofacial variation. *J. Hum. Evol.* 44, 107–32.

[18] Herrera KJ, Somarelli JA, Lowery RK, Herrera R J (2009). To what extent did Neanderthals and modern humans interact? *Biol. Rev.* 84, 245–257.

[19] Krings M, Stone A, Schmitz RW et al. (1997). Neandertal DNA sequences and the origin of modern humans. *Cell* 90, 19–30.

[20] Green R E, Krause J, Ptak S E et al. (2006). Analysis of one million base pairs of Neanderthal DNA. *Nature* 16, 330–336.

[21] Green RE, Krause J, Briggs AW et al. (2010). A draft sequence of the Neandertal genome. *Science* 328, 710–22.

[22] Serre D, Langaney A, Chech M, et al. (2004). No Evidence of Neandertal mtDNA Contribution to Early Modern Humans. *PLoS Biol.* 2, E57.

[23] Viegas J (2010). Neanderthals, humans interbred, DNA proves. *Discovery News*.

[24] Gutiérrez G, Sánchez D, Marín A (2002). A reanalysis of the ancient mitochondrial DNA sequences recovered from Neandertal bones. *Mol. Biol. Evol.* 19, 1359 66.

[25] Hebsgaard MB, Wiuf C, Gilbert MT, Glenner H, Willerslev E (2007). Evaluating Neanderthal genetics and phylogeny. *J. Mol. Evol.* 64, 50–60.

[26] Mellars P, Jennifer CF (2011). Tenfold Population Increase in Western Europe at the Neandertal – to – Modern Human Transition. *Science* 333, 623–627.

[27] Brown TA (2010). Human evolution: Stranger from Siberia. *Nature* 464, 838–839.

[28] Krause J, Fu Q, Good JM et al. (2010). The complete mitochondrial DNA genome of an unknown hominin from southern Siberia. *Nature* 464, 894–897.

[29] Reich D, Green RE, Kircher M, et al. (2010). "Genetic history of an archaic hominin group from Denisova Cave in Siberia." *Nature* 468, 1053–60.

[30] Katsnelson A (2010). New hominin found via mtDNA. *The Scientist*.

[31] Bokma F, van den Brink V, Stadler T (2012). Unexpectedly many extinct Hominins. *Evolution* 66, 2969–74.

[32] Martinón-Torres M, Dennell R, José María Bermúdez de Castro (2011). The Denisova hominin need not be an out of Africa story. *J. Hum. Evol.* 60, 251–255.

[33] Reich D, Patterson N, Kircher M, et al. (2011). Denisova admixture and the first modern human dispersals into Southeast Asia and Oceania. *Am. J. Hum. Genet.*89, 516–28.

[34] Abi-Rached L, Jobin M J, Kulkarni S, McWhinnie A Dalva et al. (2011). The shaping of modern human immune systems by multiregional admixture with archaic humans. *Science* 334, 89–94.

[35] Brown P, Sutikna T, Morwood MJ et al. (2004). A new small-bodied hominin from the Late Pleistocene of Flores, Indonesia. *Nature* 431, 1055–61.

[36] Argue D, Donlon D, Groves C, Wright R (2006). Homo floresiensis: microcephalic, pygmoid, Australopithecus, or Homo? *J. Hum. Evol.* 51, 360–74.

[37] Martin RD, Maclarnon AM, Phillips JL, Dobyns WB (2006). Flores hominid: new species or microcephalic dwarf? *The anatomical record. Part A, Discoveries in molecular, cellular, and evolutionary biology* 288, 1123–45.

[38] Ambrose SH (1998). Late Pleistocene human population bottlenecks, volcanic winter, and differentiation of modern humans. *J. Hum. Evol.* 34, 623–651.

[39] Ambrose SH (2005). Volcanic Winter, and Differentiation of Modern Humans. *Bradshaw Foundation.*

[40] Leakey R (1994). *The Origin of* Science Masters Series. New York, NY: Basic Books. pp. 87–89.

[41] Stringer CB, Andrews P (1988). Genetic and fossil evidence for the origin of modern humans. *Science* 239, 1263–8.

[42] Wolpoff MH, Spuhler JN, Smith FH, et al. (1988). Modern Human Origins. *Science* 241, 772–4.

[43] Wolpoff MH, Hawks J, Caspari R (2000). Multiregional, not multiple origins. *Am J Phys Anthropol* 112, 129–36.

[44] M'charek A (2005). *The Human Genome Diversity Project: an ethnography of scientific practice*. Cambridge University Press. p. 96.

[45] Trent RJ (2005). *Molecular medicine: an introductory text.* Academic Press. pp. 6

[46] DeSalle R, Tattersall, I (2008). *Human bones and genomes tell us about ourselves*. Texas A&M University Press. p. 146.

[47] Webster D, Wells S (2010). *Meeting the Family: One Man's Journey Through His Human Ancestry*. National Geographic Books. p. 53.

[48] Speicher MR, Stylianos E, Antonarakis A, Motulsky G (2010). *Vogel and Motulsky's Human Genetics: Problems and Approaches*. Springer. p. 606.

[49] Kutty (2009). *Adam's Gene and the Mitochondrial Eve*. Xlibris Corporation. p. 40.

[50] Cann RL, Stoneking M, Wilson AC, (1987). Mitochondrial DNA and human evolution. *Nature* 325, 31–6.
[51] Gill V (2009). Africa's genetic clocked. *BBC News*.
[52] Jorde LB, Bamshad M, Rogers AR (1998). Using mitochondrial and nuclear DNA markers to reconstruct human evolution. *BioEssays* 20, 126–36.
[53] Wall JD, Lohmueller KE, Plagnol V (2009). Detecting Ancient Admixture and Estimating Demographic Parameters in Multiple Human Populations. *Mol. Biol. Evol.* 26, 1823–7.
[54] Macaulay V, Hill C, Achilli A, et al. (2005). Single, Rapid Coastal Settlement of Asia Revealed by Analysis of Complete Mitochondrial Genomes. *Science* 308, 1034–6.
[55] Reed DL, Smith VS, Hammond SL, et al. (2004). Genetic Analysis of Lice Supports Direct Contact between Modern and Archaic Humans. *PLoS Biol* 2, e340.

Chapter 3

ANATOMY AND MORPHOLOGY OF HUMAN EVOLUTION

ABSTRACT

The Miocene Epoch was characterized by major global climatic changes that led to more seasonal conditions with increasingly colder winters north of the Equator. By the Late Miocene, in many regions inhabited by apelike primates, evergreen broad-leaved forests were replaced by open woodlands, shrublands, grasslands, and mosaic habitats, sometimes with denser-canopied forests bordering lakes, rivers, and streams. Such diverse environments stimulated novel adaptations involving locomotion in many types of animals, including primates. Our ancestors were not strictly or even heavily carnivorous. Instead, a diet that relied on tough, abrasive vegetation, including seeds, stems, nuts, fruits, leaves, and tubers, is suggested by primate remains bearing large premolar and molar teeth with thick enamel. Behaviour and morphology associated with locomotion also responded to the shift from arboreal to terrestrial life.

The development of bipedalism enabled hominins to establish new niches in forests, closed woodlands, open woodlands, and even more open areas over a span of at least 4.5 million years. Indeed, obligate terrestrial bipedalism (that is, the ability and necessity of walking only on the lower limbs) is the defining trait required for classification in the human tribe, Hominini.

3.1. Beginning of Changes in the Miocene

It is generally agreed that the taproot of the human family shrub is to be found among ape-like species of the Middle Miocene Epoch (16.4 to 11.2 million years ago [mya]) or Late Miocene Epoch (11.2 to 5.3 mya). Genetic data based on molecular clock estimates support a Late Miocene ancestry. Various Eurasian and African Miocene primates have been advocated as possible ancestors to the early hominins, which came on the scene during the Pliocene Epoch (5.3 to 2.6 mya). Though there is no consensus among experts, the primates suggested include *Kenyapithecus*, *Griphopithecus*, *Dryopithecus*, *Graecopithecus* (*Ouranopithecus*), *Samburupithecus*, *Sahelanthropus*, and *Orrorin*. *Kenyapithecus* inhabited Kenya and *Griphopithecus* lived in central Europe and Turkey from about 16 to 14 mya. *Dryopithecus* is best known from western and central Europe, where it lived from 13 to possibly 8 mya. *Graecopithecus* lived in northern and southern Greece about 9 mya, at roughly the same time as *Samburupithecus* in northern Kenya. *Sahelanthropus* inhabited Chad between 7 and 6 million years ago. *Orrorin* was from central Kenya 6 mya. Among these, the most likely ancestor of great apes and humans may be either *Kenyapithecus* or *Griphopithecus* [1, 2].

Among evolutionary models that stress the Eurasian species, some consider *Graecopithecus* to be ancestral only to the human lineage, containing *Australopithecus*, *Paranthropus*, and *Homo*, whereas others entertain the possibility that *Graecopithecus* is close to the great-ape ancestry of *Pan* (chimpanzees and bonobos) and *Gorilla* as well. In the former model, *Dryopithecus* is ancestral to *Pan* and *Gorilla*. On the other hand, others would have *Dryopithecus* ancestral to *Pan* and *Australopithecus*on the way to *Homo*, with *Graecopithecus* ancestral to *Gorilla*. This morphology-based model mirrors results of some molecular studies, which show chimpanzees, bonobos, and humans to be more closely related to one another than any of them is to gorillas; orangutans are more distantly related [3].

In a phylogenetic model that emphasizes African Miocene species, *Samburupithecus* is ancestral to *Australopithecus*, *Paranthropus*, and *Orrorin*, and *Orrorin* begets *Australopithecus afarensis*, which is ancestral to *Homo*.

The Miocene Epoch was characterized by major global climatic changes that led to more seasonal conditions with increasingly colder winters north of the Equator. By the Late Miocene, in many regions inhabited by apelike primates, evergreen broad-leaved forests were replaced by open woodlands, shrublands, grasslands, and mosaic habitats, sometimes with denser-canopied forests bordering lakes, rivers, and streams. Such diverse environments stimulated novel adaptations involving locomotion in many types of animals, including primates. In addition, there were a larger variety and greater numbers of antelope, pigs, monkeys, giraffes, elephants, and other animals for adventurous hominins to scavenge and perhaps kill. But large cats, dogs, and hyenas also flourished in the new environments; they not only would provide meat for scavenging hominins but also would compete with and probably prey upon them. In any case, our ancestors were not strictly or even heavily carnivorous. Instead, a diet that relied on tough, abrasive vegetation, including seeds, stems, nuts, fruits, leaves, and tubers, is suggested by primate remains bearing large premolar and molar teeth with thick enamel [3].

Behaviour and morphology associated with locomotion also responded to the shift from arboreal to terrestrial life. The development of bipedalism enabled hominins to establish new niches in forests, closed woodlands, open woodlands, and even more open areas over a span of at least 4.5 million years. Indeed, obligate terrestrial bipedalism (that is, the ability and necessity of walking only on the lower limbs) is the defining trait required for classification in the human tribe, Hominini [2].

3.2. Anatomical Changes during Pliocene

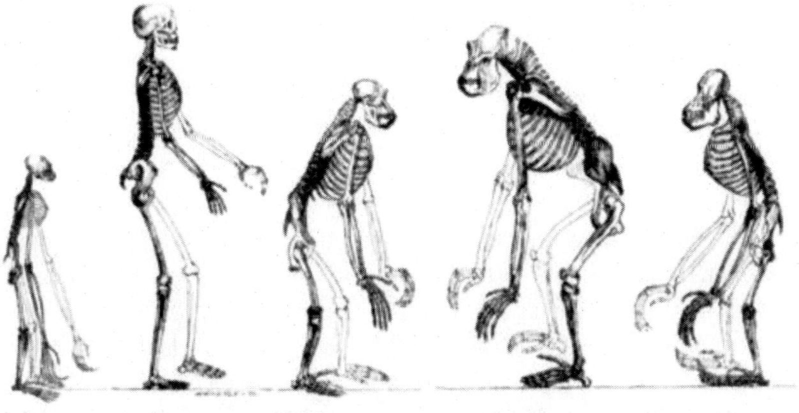

Figure 1. The hominoids are descendants of a common ancestor (Adapted from Tim Vickers 2007, en. wikipedia).

Human evolution is characterized by a number of morphological, developmental, physiological, and behavioral changes that have taken place since the split between the last common ancestor of humans and chimpanzees (Figure 1). The most significant of these adaptations are bipedalism, increased brain size, lengthened ontogeny (gestation and infancy), and decreased sexual dimorphism. The relationship between these changes is the subject of ongoing debate [4]. Other significant morphological changes included the evolution of a power and precision grip, a change first occurring in *H. erectus* [5].

3.3. The Anatomy of Bipedalism

Bipedalism is not unique to humans, though our particular form of it is. Whereas most other mammalian bipeds hop or waddle, we stride. *Homo sapiens* are the only mammal that is adapted exclusively to bipedal striding. Unlike most other mammalian orders, the primates have hind-limb-dominated locomotion. Accordingly, human bipedalism is a natural development from the basic arboreal primate body plan, in which the hind limbs are used to move about and sitting upright is common during feeding and rest [2, 4].

The initial changes toward an upright posture were probably related more to standing, reaching, and squatting than to extended periods of walking and running. Human beings stand with fully extended hip and knee joints, such that the thighbones are aligned with their respective leg bones to form continuous vertical columns. To walk, one simply tilts forward slightly and then keeps up with the displaced centre of mass, which is located within the pelvis. The large muscle masses of the human lower limbs power our locomotion and enable a person to rise from squatting and sitting postures (Figure 2).

Body mass is transferred through the pelvis, thighs, and legs to the heels, balls of the feet, and toes. Remarkably little muscular effort is expended to stand in place. Indeed, our large buttock, anterior thigh, and calf muscles are virtually unused when we stand still. Instead of muscular contraction, the human bipedal stance depends more on the way in which joints are constructed and on strategically located ligaments that hold the joints in position (Figures 2, 3). Fortunately for paleoanthropologists, some bones show dramatic signs of how a given hominin carried itself, and the adaptation to obligate terrestrial bipedalism led to notable anatomic differences between hominins and great apes. These differences are readily identified in fossils, particularly those of the pelvis and lower limbs [2, 4].

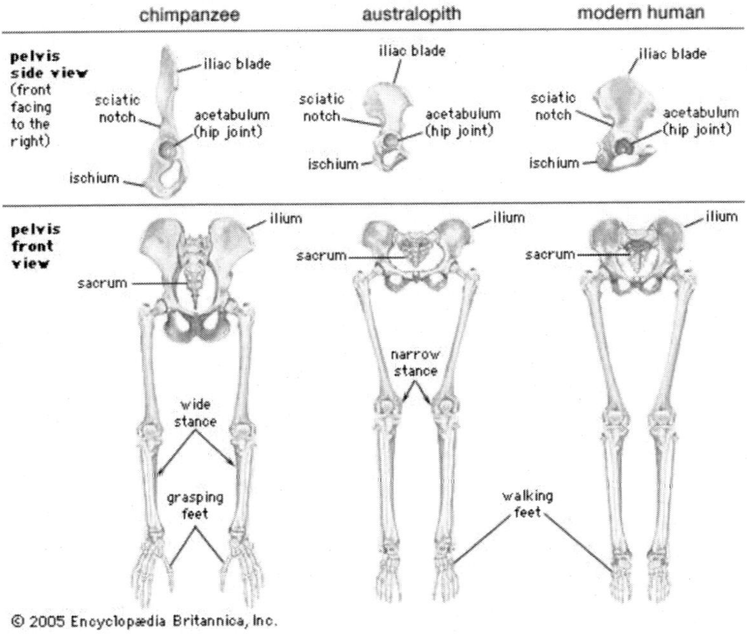

Figure 2. Comparison of the pelvis and lower limbs of a chimpanzee, an australopith, and a modern human (Adapted from Human evolution, 2015. *Encyclopedia Britannica Online*. Retrieved 18 June, 2015, from http://www.britannica.com/science/human-evolution).

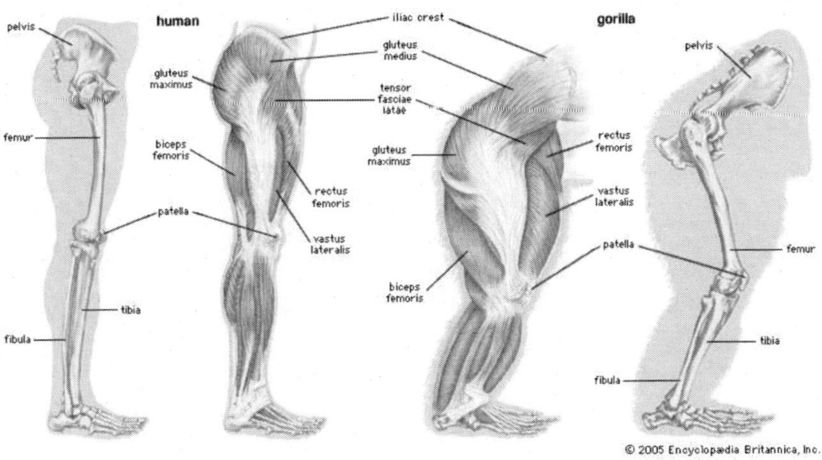

Figure 3. Skeletal and muscular structures of a human leg (left) and a gorilla leg (right), (Adapted from Human evolution, 2015. *Encyclopedia Britannica Online*. Retrieved 18 June, 2015, from http://www.britannica.com/science/human-evolution).

Although we are bipedal, our pelvis is oriented like that of quadrupedal primates. The early bipedal hominins assumed erect trunk posture by bending the spine upward, particularly in the lower back (lumbar region). In order to transfer full upper-body mass to the lower limbs and to reposition muscles so that one could walk without assistance from the upper limbs and without wobbling from side to side, changes were required in the pelvis—particularly in the ilia (the large, blade-shaped bones on either side), the ischia (protuberances on which body rests when sitting), and the sacrum (a wedge-shaped bone formed by the fusing of vertebrae) (Figures 2, 3).

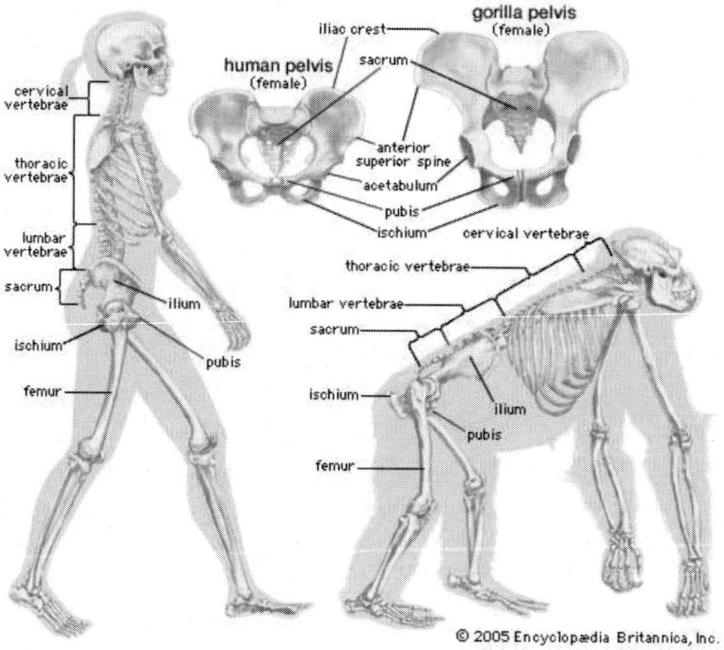

Figure 4. The skeletal structure of a human being (left) and of a gorilla (right). Several differences allow the human being to walk erect on two legs with a striding gait rather than move in a knuckle-walking fashion like the gorilla. In the pelvis these differences include shorter ischia, a broader sacrum, and broader, curved-in ilia with a lower iliac crest. In the legs the femurs (thighbones) are relatively long and are set farther apart at the hips than they are at the knees. (Adapted from Human evolution. 2015. *Encyclopedia Britannica Online*. Retrieved 18 June, 2015, from http://www.britannica.com/science/human-evolution).

Hominin hip bones have short ilia with large areas that articulate with a short, broad sacrum. Conversely, great-ape hip bones have long ilia with small sacral articular areas, and sacra of the great apes are long and narrow. The human pelvis is unique among primates in having the ilia curved forward so that the inner surfaces face one another instead of being aligned sideways, as in apes and other quadrupeds [2]. Curved ilia situate some of the gluteal muscles on the side of the hip joint, where they steady the pelvis as the foot swings forward during a step. This special mechanism allows us to walk smoothly, with only slight oscillations of the pelvis and without gross side-to-side motions of the upper body. Humans have short ischia (and long lower limbs), facilitating speedy actions of the hamstring muscles, which extend the thigh at the hip joint, while great apes have long ischia (and short hind limbs), which give them powerful hip extension for climbing up trees. Characteristically, a human thighbone is long and has a very large, globular head and a short, round neck; at the knee a prominent lateral ridge buttresses the groove in which the kneecap lies. The femurs are farther apart at the hips than at the knees and slant toward the midline to keep the knees close together. This angle allows anthropologists to diagnose bipedalism even if the fossil is only the knee end of a femur. The femurs of quadrupedal great apes, on the other hand, do not converge toward the knees, and the femoral shafts lack telltale angling as shown in figures 13 and 14 [2, 4].

Human feet are distinct from those of apes and monkeys. This is not surprising, since in humans the feet must support and propel the entire body on their own instead of sharing the load with the forelimbs. In humans the heel is very robust, and the great toe is permanently aligned with the four diminutive lateral toes. Unlike other primate feet, which have a mobile mid-foot, the human foot possesses (if not requires) a stable arch to give it strength. Accordingly, human footprints are unique and are readily distinguished from those of other animals [2] (Figure 4).

3.4. Theories of Bipedalism

There are many theories that attempt to explain why humans are bipedal, but none is wholly satisfactory. Increased speed can be ruled out immediately because humans are not very fast runners. Because bipedalism leaves the hands free, some scientists, including Darwin, linked it to tool use, especially tools for defense and hunting—i.e., weapons. This theory is problematic in that the earliest stone artifacts date only to about 2.6 mya, long after hominins had become bipedal, thus requiring an assumption that earlier tools were made of wood or other perishable materials [2, 6].

Twentieth-century theories proposed a wide array of other factors that might have driven the evolution of hominin bipedalism: carrying objects, wading to forage aquatic foods and to avoid shoreline predators, vigilantly standing in tall grass, presenting phallic or other sexual display, following migrant herds on the savanna, and conserving energy (bipedalism expends less energy than quadrupedism). Furthermore, if the early bipeds were regularly exposed to direct midday tropical sunlight, they would benefit from standing upright in two ways: less body surface would be exposed to damaging solar rays, and they would find relief in the cooler air above the ground [2, 7].

3.4.1. Advantages Associated with Bipedalism

Some scientists assume that the pre-bipedal primates were terrestrial quadrupeds, perhaps even knuckle-walkers like modern-day chimpanzees, bonobos, and gorillas. Conversely, it is also possible that the first habitual walkers were already well prepared for terrestrial bipedality, having adaptations for running bipedally among branches and boughs, standing upright to forage overhead, and climbing vertical tree trunks and vines. This scenario is suggested by studies of gibbons, which routinely engage in these arboreal activities and virtually never elect to move on the forest floor but, if forced to the ground, run bipedally [2, 7]. Gibbons have relatively long, powerful lower limbs, the same number of lumbar vertebrae that humans have (great apes have fewer), and chests of humanoid configuration. When walking on the ground, gibbons stand up straighter than chimpanzees, which are occasionally bipedal. Moreover, they exert less energy running on the ground than when running bipedally along branches or climbing vertically. Adopting a bipedal stance with full extension of the lower limbs thus would not have been a major challenge, since all apes have this capacity, though there would have been some alteration of the lower limb bones, joints, and ligaments. The foot would probably have gone through the most dramatic change, from a prehensile organ to a heel-supported, propellent one. Increased size and frequent, sustained erect standing on extended lower limbs in order to forage overhanging branches in woodland, thicket, forest edge, and other relatively open habitats would favour the evolution of humanoid hip, knee, and foot structure [2, 7]. While consuming their harvests, bipedal foragers may have squatted often, thereby further selecting for robust heels and for weight distribution between the heel and forefoot and between closely placed feet. Frequent squatting and rising would enhance development of the hamstring, buttock, and anterior thigh muscles (as hip and knee extensors), which are vital for athletic bipedalism. Stretching upward would select for shorter toes and an arched foot. Refinement of the terrestrial bipedal complex probably did not occur until hominins became less dependent upon trees for daytime refuge and other activities and began to forage widely afoot and perhaps to trek seasonally over long distances [2, 8, 9].

3.4.2. Effect of Body Size and Weight on Locomotor Efficiency

Simply increasing body size would increase locomotor efficiency, because larger animals can more effectively use the elastic energy of tendons and muscles, and they also take fewer strides to cover a given distance than a smaller animal would. Indeed, *H. rudolfensis* (2.4–1.6 mya), *H. ergaster* (1.9–1.7 mya), and later species of *Homo*, including *Homo sapiens* (100 kya), are notably taller and heavier than *Australopithecus* and *Paranthropus*. There is less size difference between the sexes in *Homo* species than in many other primates, largely because the females have become larger. Average size in male *Australopithecus* (41–51 kg [90–112 pounds]) and *Paranthropus* (40–49 kg [88–108 pounds]) is comparable to that of male chimpanzees (49 kg). The size of females (30–33, 32–34, and 41 kg, respectively) indicates that there was more difference between the sexes (sexual dimorphism) in these hominins than there is in chimpanzees [8]. Sexual dimorphism in *H. rudolfensis* (60 versus 51 kg [132 versus 112 pounds]) and *H. ergaster* (66 versus 56 kg [145 versus 123 pounds]) is comparable to that in *Homo sapiens* (58 versus 49 kg [128 versus 108 pounds]) [2].

Homo rudolfensis and *H. ergaster* (1.9–1.5 mya) have long femurs of modern human configuration and internal knee structure like that of *Homo sapiens*; both structures are quite unlike those of chimpanzees and at least some of the smaller tree-climbing primates. This may have been the time also when the distinctive morphology of the human calf muscle (triceps surae) evolved. Unlike those of great apes, it is heavily tendinous, which facilitates its function as an energy-conservant spring during walking and running [9].

3.4.3. Thermoregulation by Sweat Glands

The unique epidermal and respiratory mechanisms of *Homo sapiens* may also have developed in conjunction with regular trekking, sprinting, and endurance running as ancestral *Homo* secured a foothold in open tropical and subtropical environments.

There is a rich concentration of sweat glands in our scalp (apes have few or none in theirs), which helps to cool the head, especially the brain, in high temperatures and during vigorous activity. Post-cranially, our abundantly vascular and highly sensitive sparsely haired skin is profusely endowed with sweat glands, whose copious secretions cool an extensive surface by evaporation [10].

The distribution of sweat glands is especially strategic for cooling us while running: there is a greater concentration of sweat glands on the front surfaces of the torso and limbs, against which the air passes as we move forward. Consequently, unlike hairy quadrupeds, we do not have to pause to pant in order to avoid overheating. Furthermore, unlike the chests of quadrupeds, those of humans are freed from the stresses of supporting body weight, necessarily coupled with exhalation in running quadrupeds. We can therefore alter our breathing patterns while moving at various speeds, thereby regulating energy expenditure [11].

Homo ergaster (1.9–1.5 mya), an African species, is the earliest hominin documented with a human thoracic shape. (This species is classified by some paleoanthropologists as an African subgroup of *H. erectus*.) The thorax of Neanderthals (*H. neanderthalensis*) is also essentially like that of *Homo sapiens*, but those of other species of *Homo* are not known [12].

3.5. Refinements in Hand Structure

Primates are hand-to-mouth feeders that pluck and catch items selectively by hand before ingesting them. Without tools, emergent hominins would have relied on the versatility and strength of their hands to collect food and on their teeth and jaws alone to process it. Unless they used tools to fashion carrying devices such as bags from animal skins, they would have needed a reliable source of water nearby, and they would also have been limited in the types and number of objects that they could transport through their range (Figure 5).

In addition to transporting objects and water, there is the more obvious utility of animal skins in protecting against night chills, rain, and strong sunshine [2, 13, 14].

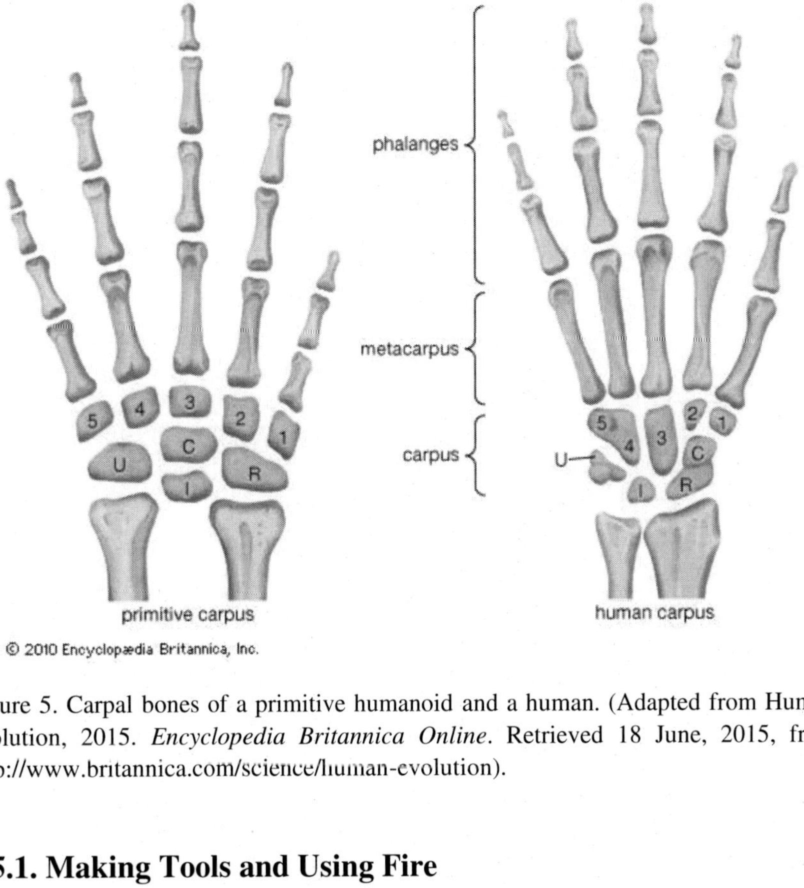

Figure 5. Carpal bones of a primitive humanoid and a human. (Adapted from Human evolution, 2015. *Encyclopedia Britannica Online*. Retrieved 18 June, 2015, from http://www.britannica.com/science/human-evolution).

3.5.1. Making Tools and Using Fire

Sharp-edged stones, even small flakes, would be a boon to early hominins who learned how to select and make them for cutting hides, meat, sticks, and other plant material. Stones also would assist in pounding open hard-shelled fruits and nuts, bones for marrow, and skulls for brains. There may have been a span when early hominins used naturally occurring stones and other objects as tools and weapons, much as some wild chimpanzees do today [2, 14].

Before hominins controlled fire and either built sturdy shelters on the ground or effectively defended caves and rock shelters, they may have constructed platforms in trees for daily activities as well as night lodging. Raw materials, stone hammers, cutting tools, and sticks and stones for defense could be stored in the trees to be used repeatedly. Handheld rocks, clubs, and long stabbing sticks, spears, or other missiles would constitute a formidable defense, especially if employed from the vantage of a tree platform [2, 15].

By about 2.6 million years ago, some hominins were making and using simple stone artifacts in eastern Africa. A likely candidate for this practice is *H. habilis*, though its contemporaries *P. boisei* and *A. garhi* cannot be overruled for this distinction. Indeed, at Bouri, Ethiopia, mammalian bones that were cut and pounded by stone tools occur in 2.5-million-year-old sediments contemporaneous with those yielding *A. garhi*.

Because the earliest stone artifacts were of such simple construction and because chimpanzees, orangutans, and capuchin monkeys today can employ stones, stems, vines, and sticks to extract nutritious morsels from protective covers, one need not expect that early hominin toolmakers displayed modern hand structure and exquisite motor control. Nonetheless, the unique structure of the human hand is readily explained by a substantial history of producing and using increasingly complex tool kits and other artifacts [2, 15]. (Attributing specific advancements in artifact manufacture to the threefold increase in brain size between Pliocene hominins and *Homo sapiens* is a much more difficult hypothesis to support)

3.5.2. Distinct Hand Anatomy from Great Apes

The features of human hands are easily distinguishable from those of the great apes, and they underpin our refined manipulatory abilities. The most complex adaptations of the human hand involve the thumb, wherein a unique, fully independent muscle (the flexor pollicis longus) gives this digit remarkable strength in pinch and power grips. The fingertips are broad and equipped with highly sensitive pads of skin. The proportional lengths of the thumb and other fingers give us an opposable thumb with precise, firm contact between its tip and the ends of each of the other fingers [15, 16]. A special saddle joint and associated ligament at the base of the thumb facilitates refined rotation. Special configurations of joints at the bases of the fifth, fourth, and second fingers facilitate tip-to-tip precision grips with the thumb. Asymmetry of the heads of the second and fifth palm bones induces rotation of the articulated fingers during opposition with the thumb. Finally, numerous modifications of the small muscles in the hand are associated with fine control of the thumb and fingers (Figure 5).

3.5.3. Australopithecus and Homo *habilis*, the Earliest Tool Maker

Australopithecus afarensis is the earliest hominin species for which there are sufficient fossil hand bones to assess manipulatory capabilities. They were capable of gripping sticks and stones firmly for vigorous pounding and throwing, but they lacked a fully developed human power grip that would allow cylindrical objects to be held between the partly flexed fingers and the palm, with counter pressure being applied by the thumb. There are insufficient specimens to assess fine manipulation in *Australopithecus*, but there is no reason to believe that they were less capable than modern chimpanzees. Chimpanzees and other apes have remarkable precision of grip, even though the tapered thumb tip must be pressed against the side of the index finger and cannot be apposed securely to any of the fingertips [15, 16].

Hand bones assigned to a 1.8-million-year-old specimen of *H. habilis* from Olduvai Gorge in northern Tanzania represent an advance over those of *A. afarensis* in features related to tool use. Tools similar to those found at Olduvai are found associated with *H. habilis* from other parts of eastern Africa as well. The tips of its thumb and fingers were flat, and there is evidence for a strong flexor pollicis longus muscle and a saddle joint at the base of the thumb. Hand bones arguably assigned to *P. robustus* or *Homo* from Swartkrans, South Africa; confirm that by about 1.8 mya one or more hominin species had highly developed thumbs and flat fingertips.

Hominin hand bones from 2.8–2.5-million-year-old cave deposits at Sterkfontein, South Africa, may be evidence that the hands of *A. africanus* were somewhat more advanced for stone tool use, but no artifact has been found in association with them [16]. Younger Sterkfontein deposits (2.0–1.5 mya) contain stone artifacts and remains of a *Homo* species.

3.5.4. Lack of *Fossil* Evidence

Because of an absence of fossils, it is not possible to track certain refinements in hand structure that must have evolved in conjunction with innovations in tool manufacture and use during the heydays of *H. rudolfensis*, *H. ergaster* (1.9–1.5 mya), and *H. erectus* (1.7–0.2 mya), as well as *H. antecessor* (1.0–0.8 mya) and *H. heidelbergensis* (600–200 kya). Only prehistoric and modern *Homo sapiens* and *H. neanderthalensis* are fully represented by hand skeletons [14, 18].

The use of tools has been interpreted as a sign of intelligence, and it has been theorized that tool use may have stimulated certain aspects of human evolution, especially the continued expansion of the human brain. Paleontology has yet to explain the expansion of this organ over millions of years despite being extremely demanding in terms of energy consumption. The brain of a modern human consumes about 13 watts (260 kilocalories per day), a fifth of body's total energy consumption [15]. Increased tool use would allow hunting for energy-rich meat products, and would enable processing more energy-rich plant products. Researchers have suggested that early hominids were thus under evolutionary pressure to increase their capacity to create and use tools [16].

Precisely when early humans started to use tools is difficult to determine, because the more primitive these tools are (for example, sharp-edged stones) the more difficult it is to decide whether they are natural objects or human artifacts. There is some evidence that the australopithecines (4 Ma) may have used broken bones as tools, but this is debated [18].

3.5.5. Oldowan Stone Tools

It should be noted that many species make and use tools, but it is the human genus that dominates the areas of making and using more complex tools. The oldest known tools are the "Oldowan stone tools" from Ethiopia, 2.5-2.6 million years old, which predates the earliest known "Homo" species. There is no known evidence that any "Homo" specimens appeared by 2.5 Ma. A *Homo* fossil was found near some Oldowan tools, and its age was noted at 2.3 million years old, suggesting that maybe the *Homo* species did indeed create and use these tools. It is a possibility but does not yet represent solid evidence [17]. Third metacarpal styloid process enables the hand bone to lock into the wrist bones, allowing for greater amounts of pressure to be applied to the wrist and hand from a grasping thumb and fingers. It allows humans the dexterity and strength to make and use complex tools. This unique anatomical feature separates humans from apes and other nonhuman primates, and is not seen in human fossils older than 1.8 million years [19].

Bernard Wood noted that "Paranthropus" co-existed with the early *Homo* species in the area of the "Oldowan Industrial Complex" over roughly the same span of time. Although there is no direct evidence which identifies *Paranthropus* as the tool makers, their anatomy lends to indirect evidence of their capabilities in this area. Most paleoanthropologists agree that the early

"Homo" species were indeed responsible for most of the Oldowan tools found. They argue that when most of the Oldowan tools were found in association with human fossils, *Homo* was always present, but *Paranthropus* was not [17].

In 1994 Randall Susman used the anatomy of opposable thumbs as the basis for his argument that both the *Homo* and *Paranthropus* species were toolmakers. He compared bones and muscles of human and chimpanzee thumbs, finding that humans have 3 muscles which are lacking in chimpanzees. Humans also have thicker metacarpals with broader heads, allowing more precise grasping than the chimpanzee hand can perform. Susman posited that modern anatomy of the human thumb is an evolutionary response to the requirements associated with making and handling tools and that both species were indeed toolmakers [17, 20].

3.6. Encephalization

The human species developed a much larger brain than that of other primates – typically 1,330 cm^3 in modern humans, over twice the size of that of a chimpanzee or gorilla [21]. The pattern of encephalization started with *Homo habilis*, which at approximately 600 cm^3 had a brain slightly larger than that of chimpanzees, and continued with *Homo erectus*(800–1,100 cm^3), reaching a maximum in Neanderthals with an average size of (1,200–1,900 cm^3), larger even than *Homo sapiens*. The pattern of human postnatal brain growth differs from that of other apes (heterochrony) and allows for extended periods of social learning and language acquisition in juvenile humans. However, the differences between the structure of human brains and those of other apes may be even more significant than differences in size [22-24].

The increase in volume over time has affected areas within the brain unequally – the temporal lobes, which contain centers for language processing, have increased disproportionately, as has the prefrontal cortex which has been related to complex decision-making and moderating social behavior [21]. Encephalization has been tied to an increasing emphasis on meat in the diet, or with the development of cooking [25], and it has been proposed that intelligence increased as a response to an increased necessity for solving social problems as human society became more complex.

3.7. Increasing Brain Size

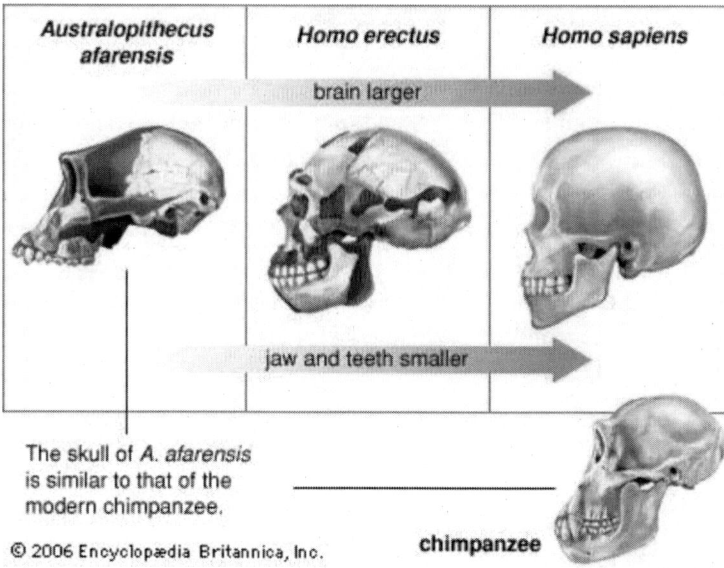

Figure 6. Evolution of relative cranial capacity and dentition patterns in selected hominins (Adapted from Human evolution, 2015. *Encyclopedia Britannica Online*. Retrieved 18 June, 2015, from http://www.britannica.com/science/human-evolution).

Because more complete fossil heads than hands are available, it is easier to model increased brain size in parallel with the rich record of artifacts from the Paleolithic Period (*c.* 2,500,000 to 10,000 years ago), popularly known as the Old Stone Age. The Paleolithic preceded the Middle Stone Age, or Mesolithic Period; this nomenclature sometimes causes confusion, as the Paleolithic itself is divided into Early, Middle, and Late (or Upper) periods. Hominin brain expansion tracks so closely with refinements in tool technology that some scholars ignore other factors that may have contributed to the brain's increasing size, such as social complexity, foraging strategies, symbolic communication, and capabilities for other culture-mediated behaviours that left no or few archaeological traces [2, 26].

Throughout human evolution, the brain has continued to expand (Figures 6, 7). Estimated average brain masses of *A. afarensis* (435 grams [0.96 pound]), *A. garhi* (445 grams [0.98 pound]), *A. africanus* (450 grams [0.99 pound]), *P. boisei* (515 grams [1.13 pounds]), and *P. robustus* (525 grams [1.16 pounds]) are close to those of chimpanzees (395 grams [0.87 pound]) and gorillas (490 grams [1.08 pounds]). Average brain mass of *Homo sapiens* is 1,350 grams (2.97 pounds). The increase appears to have begun with *H. habilis* (600 grams [1.32 pounds]), which is also notable for having a small body as given in table 1, [26]. The trend in brain enlargement continued in Africa with larger-bodied *H. rudolfensis* (735 grams [1.62 pounds]) and especially *H. ergaster* (850 grams [1.87 pounds]) [2].

One must be extremely cautious about ascribing greater cognitive capabilities, however. Relative to estimated body mass, *H. habilis* is actually "brainier" than *H. rudolfensis* and *H. ergaster*. A similar interpretive challenge is presented by Neanderthals versus modern humans. Neanderthals had larger brains than earlier *Homo* species, indeed rivaling those of modern humans. Relative to body mass, however, Neanderthals are less brainy than anatomically modern humans. Relative brain size of *Homo* did not change from 1.8 to 0.6 mya. After about 600 kya it increased until about 35,000 years ago, when it began to decrease. Worldwide, average body size also decreased in *Homo sapiens* from 35,000 years ago until very recently, when economically advanced peoples began to grow larger while less-privileged peoples did not [2, 27].

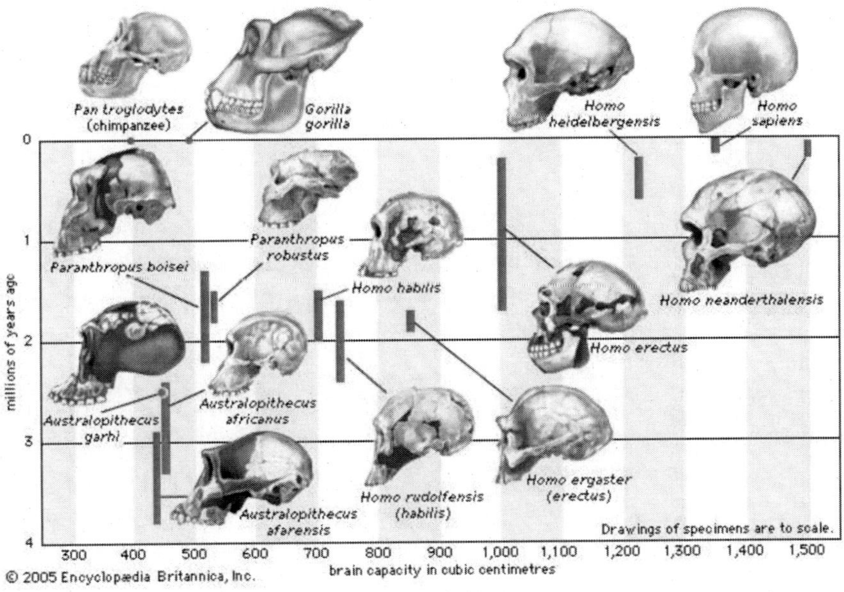

Figure 7. The increase in hominin cranial capacity over time (Adapted from Human evolution, 2015. *Encyclopedia Britannica Online*. Retrieved 18 June, 2015, from http://www.britannica.com/science/human-evolution).

Table 1. Average Capacity of the Braincase in Fossil Hominins*

Hominin	Number of fossil examples	Average capacity of the braincase (cc)
Australopithecus	6	440
Paranthropus	4	519
Homo habilis	4	640
Javanese *Homo erectus* (Trinil and Sangiran)	6	930
Chinese *Homo erectus* (Peking man)	7	1,029
Homo sapiens	7	1,350

*Adapted from Tuttle (2015). *En.Britannica* [28].

Overall, there were periods of stagnation and elaboration in stone tool technology during the Paleolithic, but, because of variations over time and between locations as well as the possibility that plant materials were used instead of stone, it is impossible to link brain size with technological complexity and fully human cognitive capabilities. Moreover, in many instances it is impossible to identify assuredly the hominin species that commanded a Paleolithic industry, even when there are associated skeletal remains at the site [2].

The unreliability of brain size to predict cognitive competence and ability to survive in challenging environments is underscored by the discovery of a distinctive human sample, dubbed *Homo floresiensis*, in a limestone cave on Flores Island, Indonesia, in 2004.

The diminutive *H. floresiensis* had brains comparable in mass to those of chimpanzees and small australopiths, yet they produced a stone tool industry comparable to that of Early Pleistocene hominins and survived among giant rats, dwarf elephants, and Komodo dragons from at least 38 kya to about 18 kya. If they are indeed a distinct species, they constitute yet another archaic human (in addition to *H. neanderthalensis* and perhaps *H. erectus*) that lived contemporaneously with modern humans during the Late Pleistocene [2, 27].

3.8. Reduction in Tooth Size

The combined effects of improved cutting, pounding, and grinding tools and techniques and the use of fire for cooking surely contributed to a documented reduction in the size of hominin jaws and teeth over the past 2.5 to 5 million years, but it is impossible to relate them precisely. It is not known when hominins gained control over fire or which species may have employed it thereafter for food preparation, warmth, or protection against predators. It is very difficult to discern whether a fire was deliberately produced by hominins or occurred naturally. For example, in a wildfire, burned-out tree stumps might leave circular accumulations of charcoal residue that could be mistaken for hearths, whereas campfires built by mobile hominins would leave no lasting evidence [2, 7].

Concentrations of charcoal, burned bones, seeds, and artifacts in China and France suggest that *H. erectus*, *H. heidelbergensis*, or both used fire as early as 460 kya. Certainly some Middle and Late Paleolithic peoples controlled fire, but hearths are rare until 100 kya. If claims for control of fire in South Africa 1.5 mya are confirmed, *P. robustus* or *H. ergaster* would be the first fire keepers [2].

At first glance early hominin skulls appear to be more like those of apes than humans. Whereas humans have small jaws and a large braincase, great apes have a small braincase and large jaws. In addition, the canine teeth of apes are large and pointed and project beyond the other teeth, whereas those of humans are relatively small and non-projecting. Indeed, human canines are unique in being incisor like, and the front lower premolar tooth is bicuspid. In apes and in many monkeys, however, the lower premolar is unicuspid and hones the upper canine tooth to razor sharpness.

In male *Australopithecus* and *Paranthropus* the large chewing muscles needed to power their deep, robust, jaws were attached to prominent crests on the braincase and to flaring arches of bone on the face and sides of the skull. Over time the rear teeth of *Paranthropus* increased in size while the incisors and canines shrank. Accordingly, *P. robustus* and *P. boisei* have relatively flat faces and non-protruding jaws [2, 7, 28].

Australopithecus species also had large rear teeth, but their faces were more protruding because the incisors and canines were not as reduced as those of *Paranthropus*. Over time the rear teeth progressively increased in size from *A. anamensis* to *A. africanus* and *H. habilis*, with *A. afarenis* intermediate between *A. anamensis* and the younger species of *Australopithecus*. When compared with estimated body size, the pattern of increased tooth size over time is confirmed for *Paranthropus* [2].

Tooth wear patterns in *A. afarensis* indicate that it may have stripped vegetal foods by manually pulling them across the front teeth. The robust-skulled *Paranthropus* may have eaten tougher foods than did gracile-skulled *Australopithecus*. Additionally, some paleoanthropologists believe that *Paranthropus* was vegetarian, while *A. africanus* had more meat in its diet. Dental morphology and wear patterns indicate that in South Africa *P. robustus* ate hard foods and that Kenyan *P. boisei*chewed whole pods and fruits with hard coatings and tough seeds, though they probably did not chew quantities of grass seed, leaves, or bone [2].

Unlike those of *Paranthropus* and *Australopithecus*, the teeth of *Homo* became smaller over time. *H. rudolfensis* has large rear teeth, even relative to estimated body size, but *H. ergaster* approaches the modern human condition. Concomitantly, the face of *H. rudolfensis* is more like that of *Australopithecus* than *H. ergaster*. One expects this trend to be related somehow to changes in diet or techniques of food preparation, but evidence to support this link is not available in the archaeological record [2, 28].

3.9. Sexual Dimorphism

The reduced degree of sexual dimorphism is visible primarily in the reduction of the male canine tooth relative to other ape species (except gibbons) and reduced brow ridges and general robustness of males. Another important physiological change related to sexuality in humans was the evolution of hidden estrus. Humans and bonobos are the only apes in which the female is fertile year round and in which no special signals of fertility are produced by the body (such as genital swelling during estrus).

Nonetheless, humans retain a degree of sexual dimorphism in the distribution of body hair and subcutaneous fat, and in the overall size, males being around 15% larger than females. These changes taken together have been interpreted as a result of an increased emphasis on pair bonding as a possible solution to the requirement for increased parental investment due to the prolonged infancy of offspring.

3.10. OTHER CHANGES

A number of other changes have also characterized the evolution of humans, among them an increased importance on vision rather than smell; a smaller gut; loss of body hair; evolution of sweat glands; a change in the shape of the dental arcade from being u-shaped to being parabolic; development of a chin (found in *Homo sapiens* alone), development of styloid processes; development of a descended larynx.

REFERENCES

[1] Mellars P (2006). Why did modern human populations disperse from Africa ca. 60,000 years ago? *Proc. Natl. Acad. Sc. USA* 103, 9381–6.

[2] Tuttle RH (2015). Human evolution (2015). *Encycl. Britann.Online.* Retrieved 18 June, 2015, from http://www.britannica.com/ science/human-evolution).

[3] Macaulay V, Hill C, Achilli A, et al. (2005). Single, Rapid Coastal Settlement of Asia Revealed by Analysis of Complete Mitochondrial Genomes. *Science* 308, 1034–6.

[4] Boyd R, Silk J B (2003). *How Humans Evolved.* New York, New York: Norton.

[5] Brues AM, Snow C C (1965). Physical Anthropology. *Biennial Review of Anthropology* 4, 1–39.

[6] Jones S, Martin RD, Pilbeam DR (Editors) (1994). *The Cambridge Encyclopedia of Human evolution.* Cambridge University Press.

[7] Leakey R, Lewin R (1992). *Origins Reconsidered: In Search of What Makes us Human.* Little, Brown and Company.

[8] Morwood M, van Oosterzee P (2007). *A New Human: The Startling Discovery and Strange Story of the 'Hobbits' of Flores, Indonesia.* Smithsonian Books.

[9] Roberts A (2009). *The Incredible Human Journey: The Story of how we Colonised the Planet.* Bloomsbury.

[10] Stringer C (2011). *The Origin of Our Species.* Allen Lane.

[11] Stringer C, Andrews P (2005). *The Complete World of Human Evolution.* Thames & Hudson.

[12] Stringer C, McKie R (1996). *African Exodus: The Origins of Modern Humanity.* Jonathan Cape.

[13] Weiss ML, Mann AE (1985). *'Human Biology and Behaviour: An anthropological perspective'* (4th ed.). Boston: Little Brown.

[14] Wells S (2004). *The Journey of Man : A Genetic Odyssey* (in English). New York, NY: Random House Trade Paperbacks.

[15] Jabr F (2012). Does Thinking Really Hard Burn More Calories?. Scientific American.

[16] Gibbons A (1998). Solving the Brain's Energy Crisis. *Science* 280, 1345–47.

[17] Freeman S, Herron JC (2007) . *Evol Anal* (4th ed.), Pearson Education, Inc.

[18] Howell F, Bourlière F (2007). *African Ecology and Human Evolution.* Transaction Publishers. p. 398. ISBN 978-0-202-36136-9.Houghton Mifflin Company (2006). *More word histories and mysteries: from aardvark to zombie*. Houghton Mifflin Harcourt. pp. 99.

[19] Ward CV, Tocheri MW, Plavcan JM, et al. (2013). Early Pleistocene third metacarpal from Kenya and the evolution of modern human-like hand morphology. *Proc Natl Acad Sc USA* 111, 121.

[20] Susman RL (1994). Fossil evidence for early hominid tool use. Science 265, 1570-1573.

[21] Schoenemann PT (2006). Evolution of the Size and Functional Areas of the Human Brain. *Ann. Rev. Anthropol* 35, 379–406.

[22] Bruner, E (2007). Cranial shape and size variation in human evolution: structural and functional perspectives . *Child's Nervous System* 23, 1357–1365.

[23] Leonard WR, Snodgrass JJ, Robertson ML (2007). Effects of brain evolution on human nutrition and metabolism. *Annu Rev Nutr.* 27, 311–27.

[24] Park MS, Nguyen AD, Aryan HE, et al. (2007). Evolution of the human brain: changing brain size and the fossil record. *Neurosurg* 60, 555–562.

[25] Organ, C (2011). Phylogenetic rate shifts in feeding time during the evolution of Homo. *Proc Natl Acad Sc USA* 108, 14555-14559.

[26] Flinn MV, Geary DC, Ward CV (2005). Ecological dominance, social competition, and coalitionary arms races: Why humans evolved extraordinary intelligence. *Evol Hum Behav* 26, 10-46.

[27] Hill A, Ward S (1988). Origin of the hominidae: The record of African large hominoid evolution between 14 my and 4 my. *Yearbook of Phy Anthropol* 31, 49–83.

[28] Lewin R (1987). *Bones of Contention: Controversies in the Search for Human Origins*. Penguin Books.

Chapter 4

EVIDENCES FOR HUMAN EVOLUTION

ABSTRACT

The evidence on which scientific accounts of human evolution is based comes from many fields of natural science. The main sources of knowledge about the evolutionary process has traditionally been the fossil record, but since the development of genetics beginning in the 1970s, DNA analysis has come to occupy a place of comparable importance. The studies of ontogeny, phylogeny and especially evolutionary developmental biology of both vertebrates and invertebrates offer considerable insight into the evolution of all life, including how humans evolved. The specific study of the origin and life of humans is anthropology, particularly paleoanthropology which focuses on the study of human prehistory.

INTRODUCTION

4.1. EVIDENCE FROM MOLECULAR BIOLOGY

The closest living relatives of humans are bonobos and chimpanzees (both genus *Pan*) and gorillas (genus *Gorilla*) [1, 2]. With the sequencing of both the human and chimpanzee genome, current estimates of the similarity between their DNA sequences range between 95% and 99% [3, 4]. By using the technique called the molecular clock which estimates the time required for the number of divergent mutations to accumulate between two lineages, the approximate date for the split between lineages can be calculated. The gibbons (family Hylobatidae) and orangutans (genus *Pongo*) were the first groups to split from the line leading to the humans, then gorillas followed by the chimpanzees and bonobos (Figure 1). The splitting date between human and chimpanzee lineages is placed around 4-8 million years ago during the late Miocene epoch [5-7].

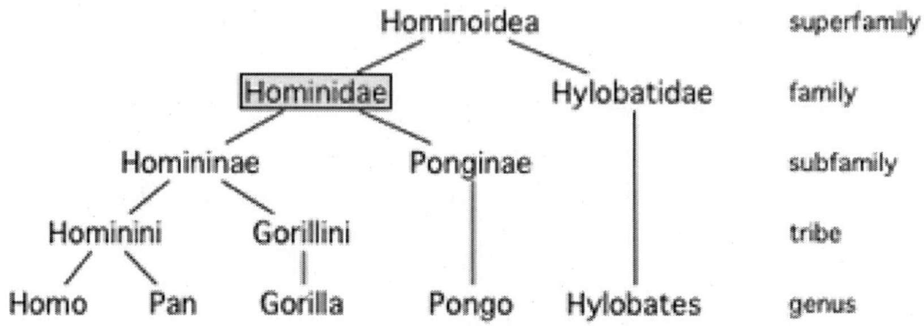

Figure 1. Family tree showing the extant hominoids: humans (genus *Homo*), chimpanzees and bonobos (genus *Pan*), gorillas (genus *Gorilla*), orangutans (genus *Pongo*), and gibbons (four genera of the family Hylobatidae: *Hylobates*, *Hoolock*, *Nomascus*, and *Symphalangus*). All except gibbons are hominids.

4.2. Genetic Evidence

Genetic evidence has also been employed to resolve the question of whether there was any gene flow between early modern humans and Neanderthals, and to enhance our understanding of the early human migration patterns and splitting dates. By comparing the parts of the genome that are not under natural selection and which therefore accumulate mutations at a fairly steady rate, it is possible to reconstruct a genetic tree incorporating the entire human species since the last shared ancestor (Figure 1).

Each time a certain mutation (Single-nucleotide polymorphism) appears in an individual and is passed on to his or her descendants a haplogroup is formed including all of the descendants of the individual who will also carry that mutation. By comparing mitochondrial DNA which is inherited only from the mother, geneticists have concluded that the last female common ancestor whose genetic marker is found in all modern humans, the so-called mitochondrial Eve, must have lived around 200,000 years ago.

4.3. EVIDENCE FROM *FOSSIL* RECORDS

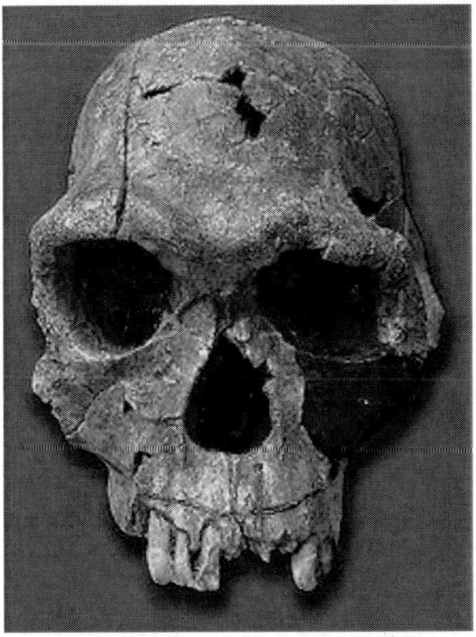

Figure 2. Replica of fossil skull of *Homo habilis*. Fossil number KNM ER 1813, found at Koobi Fora, Kenya.

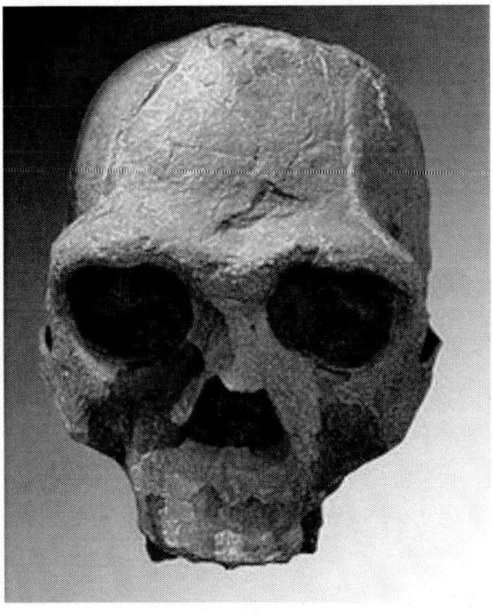

Figure 3. Replica of fossil skull of *Homo ergaster* (African *Homo erectus*). Fossil number Khm-Heu 3733 discovered in 1975 in Kenya.

There is little fossil evidence for the divergence of the gorilla, chimpanzee and hominin lineages [7]. The earliest fossils that have been proposed as members of the hominin lineage are *Sahelanthropus tchadensis* dating from 7 million years ago, *Orrorin tugenensis* dating from 5.7 million years ago and *Ardipithecus kadabba* dating to 5.6 million years ago. Each of these has been argued to be a bipedal ancestor of later hominins but, in each case, the claims have been contested. It is also possible that one or more of these species are ancestors of another branch of African apes, or that they represent a shared ancestor between hominins and other apes (Figures 2, 3).

The question of the relationship between these early fossil species and the hominin lineage is still to be resolved. From these early species, the australopithecines arose around 4 million years ago and diverged into robust (also called *Paranthropus*) and gracile branches, one of which (possibly *A. garhi*) probably went on to become ancestors of the genus *Homo*. The australopithecine species that is best represented in the fossil record is *Australopithecus afarensis* with more than one hundred fossil individuals represented, found from Northern Ethiopia (such as the famous "Lucy"), to Kenya, and South Africa. Fossils of robust australopithecines such as *A. robustus* (or alternatively *Paranthropus robustus*) and *A./P. boisei* are particularly abundant in South Africa at sites such as Kromdraai and Swartkrans, and around Lake Turkana in Kenya (Figures 2, 3).

4.3.1. Fossil Evidence for the Use of Tools

The earliest members of the genus *Homo* are *Homo habilis* which evolved around 2.3 million years ago. *Homo habilis* is the first species for which we have positive evidence of the use of stone tools. They developed the oldowan lithic technology, named after the Olduvai Gorge in which the first specimens were found. Some scientists consider *Homo rudolfensis*, a larger bodied group of fossils with similar morphology to the original *H. habilis* fossils, to be a separate species while others consider them to be part of *H. habilis*—simply representing species internal variation, or perhaps even sexual dimorphism. The brains of these early hominins were about the same size as that of a chimpanzee, and their main adaptation was bipedalism as an adaptation to terrestrial living.

4.3.2. Fossil Evidence for *encephalization*

During the next million years, a process of *encephalization* began and, with the arrival of *Homo erectus* in the fossil record, cranial capacity had doubled. *Homo erectus* was the first of the *hominina* to leave Africa, and this species spread through Africa, Asia, and Europe between 1.3 to 1.8 million years ago. One population of *H. erectus*, also sometimes classified as a separate species *Homo ergaster*, stayed in Africa and evolved into *Homo sapiens*. It is believed that these species were the first to use fire and complex tools.

4.3.3. *Fossil* Evidence for Dispersal

The earliest transitional fossils between *H. ergaster/erectus* and Archaic *H. sapiens* are from Africa, such as *Homo rhodesiensis*, but seemingly transitional forms were also found at Dmanisi, Georgia. These descendants of African *H. erectus* spread through Eurasia from ca. 500,000 years ago evolving into *H. antecessor*, *H. heidelbergensis* and *H. neanderthalensis*. The earliest fossils of anatomically modern humans are from the Middle Paleolithic, about 200,000 years ago such as the Omo remains of Ethiopia; later fossils from Skhul in Israel and Southern Europe begin around 90,000 years ago.

As modern humans spread out from Africa, they encountered other hominins such as *Homo neanderthalensis* and the so-called Denisovans, who may have evolved from populations of *Homo erectus* that had left Africa around 2 million years ago. The nature of interaction between early humans and these sister species has been a long standing source of controversy, the question being whether humans replaced these earlier species or whether they were in fact similar enough to interbreed, in which case these earlier populations may have contributed genetic material to modern humans [8, 9].

This migration out of Africa is estimated to have begun about 70,000 years BC and modern humans subsequently spread globally, replacing earlier hominins either through competition or hybridization. They inhabited Eurasia and Oceania by 40,000 years BP, and the Americas by at least 14,500 years BP [10].

4.3.4. Fossil Evidence for *Bipedalism*

By 3.5 million years ago at least one hominin species, *A. afarensis*, was an adept walker. In addition to anatomic evidence from this time, there is also a 27.5-metre (90-foot) track way produced by three individuals who walked at a leisurely pace on moist volcanic ash at Laetoli in northern Tanzania. In all observable features of foot shape and walking pattern, they are astonishingly similar to those of habitually barefoot people who live in the tropics today. Nevertheless, although the feet of the Laetoli hominins appear to be strikingly human, one should not assume that other parts of their bodies were as similar to ours [11, 12].

The fragmentary femoral remains found in Kenya of six-million-year-old *Orrorin tugenensis* indicate to some experts that they too were bipeds. *Ardipithecus ramidus* (5.8–4.4 mya), a primate from Aramis, central Ethiopia, was also bipedal. In this case the evidence comes from the foramen magnum, the hole in the skull through which the spinal cord enters. In *Ardipithecus* this opening is similar to ours in being located centrally under the skull instead of at the rear of it. A rear-facing foramen magnum indicates a stooped posture, whereas a downward-facing hole positions the skull atop the spinal column. Other characteristics indicative of bipedalism in *Ardipithecus* include an increased tarsal region in each foot and a pelvic structure with muscle-to-bone attachment sites comparable to later, bipedal hominins. In addition, the leg bone of *Australopithecus anamensis* from northern Kenya (4.2–3.9 mya) attests to its bipedalism [13].

All hominins living at the time of the Laetoli track makers were probably obligate bipeds when on the ground, but some of them (including some younger species) exhibit features that argue for regular arboreal climbing, probably for food, rest, nightly lodging, and predator avoidance. Hadar, in northern Ethiopia, has yielded a trove of remains of *A. afarensis* (3.8–2.9 mya). They include many parts of the locomotor skeleton that reveal a bipedal habit: short ilia, a wide and stout sacrum, and femoral angling, among other features. At the same time, the curved fingers and toes, laterally flared ilia, and short femurs with long upper limbs, as well as the configuration of its rib cage, indicate that they could readily climb and maneuver in trees. *A. bahrelghazali* (3.5–3.0 mya) of central Chad and *Kenyanthropus platyops* (3.5 mya) from northern Kenya are represented solely by teeth and by skull and jaw fragments from which positional behaviour cannot be inferred [14].

4.3.5. Fossil Evidence for *Locomotor Skelton*

Parts of the locomotor skeletons of later hominins such as *A. africanus* (3.3–2.4 mya) and *Paranthropus robustus* (1.8–1.5 mya) of South Africa do not differ markedly from those of *A. afarensis*. The locomotor skeleton of eastern African *P. boisei* (2.2–1.3 mya) is poorly known, but there is no reason to assume that it was different from other *Paranthropus* species. Bouri, a 2.5-million-year-old site in central Ethiopia, yielded arm and leg bones that are contemporaneous with craniodental remains of *A. garhi*. The femur is elongated relative to the humerus, as in *Homo sapiens*, but, unlike the human forearm, that of the fossil specimen is relatively long. Thus, by 2.5 mya at least one hominin species had developed the long femurs of striding bipeds, though it retained long forearms like arboreally active *Australopithecus* and *Paranthropus* [15].

Homo habilis (2.0–1.5 mya), best known from Olduvai Gorge, Tanzania, exhibits small teeth and a large brain, but it has long upper limbs (especially the forearms), short femurs, curved finger bones, and other chimpanzee-like traits that indicate a mélange of arboreal and terrestrial adaptations. Because of these similarities, some investigators classify *H. habilis* as *A. habilis*.

4.3.6. Fossil Evidence for Pelvis

The pelvis of *H. heidelbergensis* (600,000–200,000 years ago, or 600–200 kya) and that of Neanderthals(200–30 kya) are distinct from the pelvis of *Homo sapiens* in some features that recall those of *Australopithecus*. The pelvis is broad, with ilia flaring out to the side. The femoral necks are also relatively long. These features are related to stabilizing the pelvis in stocky bipedal hominins. The pelvises of both *H. heidelbergensis* and Neanderthals could accommodate a wider birth canal. This feature is important because they may have had notably larger brains (about 1,200 grams [2.65 pounds] and 1,400 grams (3.09 pounds), respectively) than earlier hominins did, a trait that is reflected in the size of the fetal skull [16].

4.3.7. Lack of Fossil Evidence for Development of Foot

Regrettably, development of foot structure in early *Homo*—i.e., between *A. afarensis* and Neanderthals—is virtually undocumented by skeletal evidence. The oldest footprints indicative of contemporary foot function, however, have been found in Ileret, Kenya. These prints have been dated at 1.51 to 1.53 mya, and their size and depth suggest that they were made by *H. ergaster* or *H. erectus*. Therefore, it is safe to assume that by about 1.53 mya the uniquely human locomotor and associated cooling systems were basically established. Subsequent alterations in pelvic shape may be related to the passage of larger-brained babies through the birth canal [17, 18].

REFERENCES

[1] Stanford C, Allen JS, Anton SC (2012). Biological Anthropology (2nd Edition). Prentice Hall.

[2] Wood B, Richmond BG (2000). Human evolution: taxonomy and paleobiology. *J Anat* 197, 19–60.

[3] Varki A, Nelson DL (2007). Genomic Comparisons of Humans and Chimpanzees. *Ann. Rev. Anthropol.* 36,191–209

[4] Sayers K, Raghanti MA, Lovejoy CO (2012). Human Evolution and the Chimpanzee Referential Doctrine. *Ann Rev Anthropol*, 41.

[5] Ruvolo M (1997). Molecular phylogeny of the hominoids: inferences from multiple independent DNA sequence data sets. *Mol Biol Evol* 14, 248–265.

[6] Dawkins R (2004). *The Ancestor's Tale*. pp 136.

[7] Begun, D R (2010). Miocene Hominids and the Origins of the African Apes and Humans. *Ann Rev Anthropol* 39, 67 -84.

[8] Wood BA (2009). Where does the genus *Homo* begin, and how would we know? In: Grine FE, Fleagle JG, Leakey RE (eds). *The First Humans: Origin and Early Evolution of the Genus* Homo. London, UK: Springer. pp. 17–27.

[9] Mitchell A (2012). DNA Turning Human Story Into a Tell-All. NYTimes.

[10] Wood B (1996). Human evolution. *BioEssays* 18, 945–954.

[11] Hartwig W ed. (2002). *The Primate Fossil Record*. Cambridge University Press.

[12] Tattersall I (2008). *The Fossil Trail: How We Know What We Think We Know About Human Evolution*. Oxford University Press.

[13] Weiss ML, Mann AE (1985). *'Human Biology and Behaviour: An anthropological perspective* (4th ed.). Boston: Little Brown.

[14] Alexander RD (1990). How did humans evolve? Reflections on the uniquely unique species. *University of Michigan Museum of Zoology Special Publication* 1, 1–38.

[15] Gibbons A (2007). *The First Human: The Race to Discover our Earliest Ancestor*. Anchor Books.

[16] Heizmann, Elmar P J, Begun, David R (2001). The oldest Eurasian hominoid. *Journal of Human Evolution* 41 5, 463–81.

[17] Hill A, Ward S (1988). Origin of the hominidae: The record of African large hominoid evolution between 14 my and 4 my. *Yearbook of Phy Anthropol* 31, 49–83.

[18] Jones S, Martin RD, Pilbeam DR (Eds.) (1994). *The Cambridge Encyclopedia of Human evolution*. Cambridge University Press.

Chapter 5

EVOLUTION OF HUMAN BEHAVIOR

ABSTRACT

This section backgrounds the beginnings in the Miocene Era and describes certain global climatic changes that reduced forested areas and induced more open terrestrial biomes during the late Miocene Epoch (11.2–5.3 mya). During the succeeding Pliocene Epoch (5.3–2.6 mya) these changes only intensified. In Africa, primates diversified. In Eurasia, contrarily, hominins disappeared by the beginning of the Pliocene. The only descendants of the Late Miocene primates in Asia were the extinct Early-Middle Pleistocene *Gigantopithecus blacki* of southern China and northern Vietnam, and the present-day orangutans and gibbons of South and Southeast Asia.

5.1. HOMININ HABITATS

It is reasonable to expect that the increased variety and shifting distribution of African biomes stimulated new hominin lifeways, some of which led to survival and others of which did not. Insofar as habitats have been (or can be) discerned from evidence found with the Pliocene hominin species, hominins inhabited a variety of biomes in eastern, central, and southern Africa. In central Ethiopia, *Ardipithecus ramidus* is associated with faunal and floral remains indicating a woodland habitat. Later remains, in northern Ethiopia, indicate *Australopithecus afarensis* inhabited a mosaic of riverine forest, lowland woodland, savanna, and dry bushland. In northern Kenya *Australopithecus anamensis* lived in dry open woodland or bushland with a gallery forest along a nearby river. In central Chad the northernmost and westernmost species, *Australopithecus bahrelghazali*, appears to have lived in a mosaic of open and wooded biomes near a river. Mammalian fossils from Lomekwi, northern Kenya, indicate that *Kenyanthropus platyops* inhabited a relatively well-watered area of forest or closed woodland or the forest edge between them. The habitat of the 3.5-million-year-old Laetoli hominins in northern Tanzania was arguably a mosaic of open grassland and more-closed woodland [1]. The area may have been wetter than it is now. No permanent water source has been identified for the Laetoli area during the Pliocene. Later in the Pliocene, *Australopithecus garhi* was active on broad, grassy plains bordering a lake in central Ethiopia. Models of the habitat of *Australopithecus africanus*, based on fauna from the two major South African cave sites—*Sterkfontein* and *Makapansgat*—stress closed-canopy wooded conditions: either dry woodland with grasslands nearby or subtropical forest. During the tenures of *H. habilis* and *P. boisei* at Olduvai Gorge, northern Tanzania, the climate changed from moist to dry and again to moist before a long dry span that began two million years ago. Specimens of both of these Olduvai hominins are mostly from the shore of an ancient saline, alkaline lake. At Koobi Fora, northern Kenya, specimens of *H. habilis* have been more commonly found in lake-margin deposits, while those of *P. boisei* are equally common in river and lake-margin sediments. Fossil pollen indicates that highland forest was nearby and that near the lake there were grassy areas and dense woodland and shrub land [1].

At Konso, southern Ethiopia, *P. boisei* lived in a grassland habitat. Elsewhere in eastern Africa, *P. aethiopicus* was associated with closed habitats. The South African cave sites (Swartkrans, Kromdraai, and Drimolen) of *P. robustus* are associated with open and even arid habitats, but these may not reflect its actual foraging preference.

One of the more profound effects of Pliocene habitat changes was honing the energy-conservant bipedal stride at the time that *Homo* species deployed out of Africa and into Eurasia. Shortly after *Homo* evolved in Africa, some species ventured to temperate biomes in Eurasia and then to subtropical and tropical biomes in South and Southeast Asia. Subsequently there was a migration back to Africa, perhaps as early as 1.8–0.9 mya. This hemispheric dispersion of *Homo* is associated with elaboration of stone tool kits, increased brain size, and reduction in size of the jaws and teeth—all of which are the subject of the next section [2].

5.2. Tools Making in the Pliocene and Pleistocene

Stone Tools

Figure 1. Acheulean hand-axes from Kent. *Homo erectus* flint work. The types shown are (clockwise from top) cordate, ficron and ovate.

Stone tools are first attested around 2.6 Ma, when *H. habilis* in Eastern Africa used so-called pebble tools, choppers made out of round pebbles that had been split by simple strikes [3]. This marks the beginning of the Paleolithic, or Old Stone Age; its end is taken to be the end of the last Ice Age, around 10,000 years ago. The Paleolithic is subdivided into the Lower Paleolithic (Early Stone Age, ending around 350,000–300,000 years ago), the Middle Paleolithic (Middle Stone Age, until 50,000–30,000 years ago), and the Upper Paleolithic.

5.3. Levallois Technique

The period from 700,000–300,000 years ago is also known as the Acheulean, when *H. ergaster* (or *erectus*) made large stone hand axes out of flint and quartzite, at first quite rough (Early Acheulian), later "retouched" by additional, more subtle strikes at the sides of the flakes (Figure 1,2). After

350,000 BP (Before Present) the more refined so-called Levallois technique was developed, a series of consecutive strikes, by which scrapers, slicers ("racloirs"), needles, and flattened needles were made [3]. Finally, after about 50,000 BP, ever more refined and specialized flint tools were made by the Neanderthals and the immigrant Cro-Magnons (knives, blades, skimmers). In this period they also started to make tools out of bone.

5.4. Refinements in Tool Design

Figure 2. Replica stone tools of the Acheulean industry, used by Homo erectus and early modern humans, and of the Mousterian industry, used by Neanderthals. (Top, left to right) Mid-Acheulean bifacial hand ax and Acheulean banded-flint hand ax. (Centre) Acheulean hand tool. (Bottom, left to right) Mousterian bifacial hand ax, scraper, and bifacial point.

In Africa the Early Paleolithic (2.5–0.2 mya) comprises several industries with the earliest man-made chipped flakes and core choppers (2.5–2.1 mya). Double-faced hand axes, cleavers, and picks (collectively known as bifaces) appeared about 1.5 mya and persisted until about 200 kya (Figure 2). Archaeologists have detected some improvements of technique and product during the half-million-year span of core-flake industries. Although the major biface industry—the Acheulean—has been characterized as basically static, it too shows evidence of refinement over time, finally resulting in elegant, symmetrical hand axes that required notable skill to make [2].

By 1.7 mya a population of *H. erectus* similar to African *H. ergaster* lived in Eurasia at what is now Dmanisi, Georgia. The associated choppers, chopping tools, flakes, and scrapers recall the Oldowan core-flake industry of eastern Africa (Figure 3), but there are no bifaces among them. The braincase of the two Dmanisi specimens is smaller than that of African *H. ergaster*. New geochemical dates for classic hominin localities in Java indicate that *H.*

erectus may have lived in Southeast Asia 1.5 mya, but no industry is certainly identified with them [4].

Figure 3. A sharp rock", an Oldowanpebble tool, the most basic of human stone tools.

El Ubeidīya, Israel, provides evidence that people and bifaces had spread out of Africa by 1.4 mya. In Europe, Acheulean tools appear 500 kya and persist until about 250–150 kya; they also occur in South Asia. Sites in China (800 kya), Korea, and Japan contain bifaces, but they differ from Acheulean tools. No such technology has been found in tropical Southeast Asia, where bamboo tools may have sufficed.

5.5. Regional Variations in Tool Making

In both Africa and Eurasia the Middle Paleolithic was long thought to have lasted from about 200 kya to as recently as 30 kya, depending upon location. While tools from the the Early Paleolithic slowly changed across space and time, the Middle Paleolithic was characterized by an explosion of local and regional variations in size and shape and by frequencies of reshaped flakes, blades, scrapers, hand axes, and other tools. Projectile points began to be emphasized in some regions, with bone being used as well as stone; bone arrow points dating to more than 60,000 years ago have been found at Sibudu Cave in South Africa [5].

5.6. Gademotta Tool Makers

Although they vary across time and space, Middle Paleolithic tools as a whole are characterized by carefully prepared cores from which elegant flakes or blades were struck. Notably, tools of this type have been found at the Gademotta site in Ethiopia's Rift Valley in stratigraphic levels that date to approximately 275 kya. This find not only challenges the common starting date for the Middle Paleolithic but also raises intriguing questions about the Gademotta tool makers; fossil remains of fully modern humans do not appear in the archaeological record until about 200 kya, suggesting that these tools were either made by another species or that fully modern humans arose considerably earlier than was once thought [6].

5.7. Late Paleolithic Tools

Late Paleolithic industries dating to 50–10 kya comprise diverse blade and microblade tools, especially in Europe. Late Paleolithic peoples used a variety of materials for their tools and bodily ornaments, including bone, stone, wood, antler, ivory, and shell. Stone blades were long, thin, and very effective cutting tools. Often, when they became dull, someone retouched them via pressure flaking, which requires fine motor control and coordination. Microblades and other points were probably hafted to produce throwing and stabbing spears. Other composite tools of the period include atlatls, harpoons, fish weirs, and bows and arrows. Late Paleolithic people also developed techniques for grinding and polishing, with which they made beads, pendants, and other artistic objects. They also made needles (perhaps for sewing fitted clothing), fish hooks, and fish gorges [7].

5.8. Transition to Behavioral Modernity

Until about 50,000–40,000 years ago the use of stone tools seems to have progressed stepwise. Each phase (H. habilis, H. ergaster, H. neanderthalensis) started at a higher level than the previous one, but after each phase started, further development was slow. Currently paleoanthropologists are debating whether these Homo species possessed some or many of the cultural and behavioral traits associated with modern humans such as language, complex symbolic thinking, technological creativity etc. It seems that they were culturally conservative maintaining simple technologies and foraging patterns over very long periods.

Around 50,000 BP modern human culture started to evolve more rapidly. The transition to behavioral modernity has been characterized as a Eurasian "Great Leap Forward" [8], or as the "Upper Palaeolithic Revolution" [9], because of the sudden appearance of distinctive signs of modern behavior in the archaeological record. Some scholars consider the transition to have been more gradual, with some features already appearing among Archaic African *Homo sapiens* already around 200,000 years ago [10].

Modern humans started burying their dead, using animal hides to make clothing, hunting with more sophisticated techniques (such as using trapping pits or driving animals off cliffs), and engaging in cave painting [11]. As human culture advanced, different populations of humans introduced novelty to existing technologies: artifacts such as fish hooks, buttons and bone needles show signs of variation among different populations of humans, something that had not been seen in human cultures prior to 50,000 BP. Typically, *H. neanderthalensis* populations do not vary in their technologies [12].

Among concrete examples of Modern human behavior, anthropologists include specialization of tools, use of jewellery and images (such as cave drawings), organization of living space, rituals (for example, burials with grave gifts), specialized hunting techniques, exploration of less hospitable geographical areas, and barter trade networks. Debate continues as to whether a "revolution" led to modern humans ("the big bang of human consciousness"), or whether the evolution was more gradual [13].

REFERENCES

[1] Stringer C, McKie R (1996). African Exodus: The Origins of Modern Humanity. Jonathan Cape.

[2] Weiss ML, Mann AE (1985). 'Human Biology and Behaviour: An anthropological perspective (4th ed.). Boston: Little Brown.

[3] Plummer T (2004). Flaked stones and old bones: Biological and cultural evolution at the dawn of technology. Am. J. Phys. Anthropol. Suppl 39, 118–64.

[4] Stringer C, Andrews P (2005). The Complete World of Human Evolution. Thames & Hudson.

[5] Shreeve J (1996). The Neanderthal Enigma: Solving the Mystery of Modern Human Origins. Viking.

[6] Alexander RD (1990). How did humans evolve? Reflections on the uniquely unique species. University of Michigan Museum of Zoology Special Publication 1, 1–38.

[7] Flinn MV, Geary DC, Ward CV (2005). Ecological dominance, social competition, and coalitionary arms races: Why humans evolved extraordinary intelligence. Evol Hum Behav 26, 10-46.

[8] Diamond J (1999). Guns, Germs, and Steel: The Fate of Human Societies. W. W. Norton. p. 39.

[9] Ofer B-Y (2002). The Upper Paleolithic revolution. Ann Rev Anthropol 31, 363-393.

[10] Nowell A (2010). Defining behavioral modernity in the context of Neanderthal and anatomically modern human. Ann Rev Anthropol 39, 437-452.

[11] Ambrose SH (2001). Paleolithic technology and human evolution. Science 291, 1748–53.

[12] d'Errico F, Stringer CB (2011). Evolution, revolution or saltation scenario for the emergence of modern cultures? Phil. Trans. R. Soc. B 366, 1060-1069.

[13] Mcbrearty S, Brooks AS (2000). The revolution that wasn't: a new interpretation of the origin of modern human behavior. J. Hum. Evol. 39, 453–563.

Chapter 6

RECENT AND CURRENT TRENDS IN HUMAN EVOLUTION

ABSTRACT

It has been argued that human evolution has accelerated since the development of agriculture and civilization some 10,000 years ago. It is claimed that this has resulted in substantial genetic differences between different current human populations. Lactase persistence is an example of such recent evolution. Recent human evolution seems largely however to have been confined to genetic resistance to some infectious disease, which have appeared in human populations by crossing the species barrier from domesticated animals [4, 5]. Without a selection pressure or opportunity for speciation, such as geographic isolation, recent human evolution has been predominantly subject to genetic drift.

INTRODUCTION

Natural selection occurs in modern human populations. For example, the population which is at risk of the severe debilitating disease kuru has significant over-representation of an immune variant of the prion protein gene G127 V versus non-immune alleles. The frequency of this genetic variant is due to the survival of immune persons [1]. Other reported evolutionary trends in other populations include a lengthening of the reproductive period, reduction in cholesterol levels, blood glucose and blood pressure [2-5].

6.1. THE EMERGENCE OF *HOMO SAPIENS*

The relationships among *Australopithecus*, *K. platyops*, *Paranthropus*, and the direct ancestors of *Homo* are unknown. Because of its early date and geographic location, *A. anamensis* may be the common ancestor of *A. afarensis*, *A. garhi*, *K. platyops*, and perhaps the Laetoli Pliocene hominins of eastern Africa, *A. bahrelghazali* of central Africa, and *A. africanus* of southern Africa. *A. afarensis* in turn may be ancestral to *P. aethiopicus*, which begat *P. boisei* in eastern Africa and *P. robustus* in southern Africa [6-12].

Factors indicating *H. rudolfensis* as ancestral to later species of *Homo* are its absolute brain size, large body, and lower limb morphology. These features clearly foreshadow younger species of *Homo* in Africa and Eurasia.

Our ancestry becomes no clearer as the candidates are narrowed to *Homo* species exclusively. Among paleoanthropologists who accept it as a species distinct from *H. erectus*, *H. ergaster* is most often proposed as the ancestor of *Homo* species of the Pleistocene Epoch. *H. heidelbergensis* may have arisen from *H. ergaster*, *H. erectus*, or *H. antecessor*, and any or none of them could have been ancestors of *H. neanderthalensis* and *Homo sapiens*. Neanderthal populations, particularly as represented by specimens from Western Europe, probably were not ancestral to modern humans [11, 12].

Theorists use fossil remains, genetic traits of modern people around the world, and archaeological and anatomical indicators of cognitive, linguistic, and technological capabilities to support their models of recent human evolution, but no single theory provides definitive resolution of how *Homo sapiens* came to be. The limitations of empirical evidence confound efforts to discern whether distinctive features and lineages developed gradually or over periods of stasis punctuated by rapid change (a theory known as punctuated equilibrium). There are claims for about 20 fossil hominin species over the course of the last six million years, but they are assessed on a case-by-case basis [9]. For example, it appears that Neanderthals (*H. neanderthalensis*) were a dead end for two ancestral species (*H. antecessor* and *H. heidelbergensis*) that changed gradually in Europe from about 700 kya to 30 kya. *Homo sapiens* may have evolved similarly through a series of species represented by African specimens, but other theorists envision a dramatic shift in cognitive capacity and behaviour that qualifies instead as a punctuational change. This change would have occurred about 200 kya in one small African population and would have been followed by a long period of stasis that continues to the present. Such a scenario is not unprecedented, as *A. afarensis* was a capable biped that appears to have emerged suddenly and persisted for nearly one million years [6-9].

6.2. Models of Human Evolution

There are four basic models that purport to explain the evolution of *Homo sapiens* between about 200 and 30 kya. At one extreme is multiregional evolution, or the regional continuity model. At the other is the African replacement, or "out of Africa," model. Intermediate are the African hybridization-and-replacement model and the assimilation model. All but the multiregional model maintain that *Homo sapiens* evolved solely in Africa between about 200 and 100 kya and then deployed to Eurasia and eventually the Americas and Oceania. Both of the replacement models argue that anatomically modern emigrants replaced resident Eurasian and Australasian species of *Homo sapiens* with little or no hybridization [9-11]. The hybridization-and-replacement model proposes some interbreeding with archaic indigenous populations but with relatively minor effects. Assimilation maintains continuity between archaic and modern humans, most notably in some areas of Eurasia, where gene flow and local selective factors would also produce morphological changes. In this model, unity of the species was maintained by periodic interbreeding across wide areas. Multi-regionalists reject the idea that *Homo sapiens* evolved uniquely in Africa. Instead, they advocate that discrete archaic populations of *Homo* evolved locally in Africa, Asia, and Europe. Throughout their tenures, both the archaic and descendant populations interbred with contemporaries from other areas [9-11].

The African replacement model has gained the widest acceptance owing mainly to genetic data (particularly mitochondrial DNA) from existing populations. This model is consistent with the realization that modern humans cannot be classified into subspecies or races, and it recognizes that all populations of present-day humans share the same potential.

6.3. Bipedalism and Nomadism

A tangled line of descent is not surprising given the nomadic lifestyles enabled by bipedalism. There appear to have been successive migrations of hominin species out of Africa, with evolution of new species in Eurasia and occasional migrations back into Africa. For instance, *H. ergaster* may have been the first hominin to reach Eurasia. Some of its descendants could have moved quickly to East and Southeast Asia, where they begat *H. erectus*. Others may have evolved into *H. heidelbergensis*, which populated Europe sparsely and then returned to Africa [7, 8].

Some paleoanthropologists claim that *H. antecessor*, found in 800,000-year-old cave deposits at Gran Dolina, Sierra de Atapuerca, Spain, was a direct ancestor of *H. neanderthalensis* via *H. heidelbergensis*, which is represented by 300,000-year-old specimens from Sima de los Huesos in the Sierra de Atapuerca. Further, they propose that *H. antecessor*, from million-year-old deposits in Eritrea, is a direct ancestor of *Homo sapiens* in Africa.

6.4. Evolution of Neanderthals

Neanderthals probably evolved in Europe at least partially in response to cold climatic conditions and then migrated to western Asia, where they may have encountered *Homo sapiens* in the Levant. There is no skeletal evidence that they reached the African continent or moved much farther east than Uzbekistan in Central Asia. Features of Neanderthals that argue for adaptation to seasonally frigid biomes include stocky torsos, short limbs (particularly the forearms and legs), and distinctive facial structure [7, 8]. The middle of the face protrudes, the teeth are set forward, the enlarged cheekbones sweep backward, and the nasal passages are voluminous. If Neanderthals wore animal furs and other insulating materials on their heads and bodies while keeping vigorously active in frigid weather, the large nasal chamber would help to cool the blood and prevent overheating the brain, while clothing would reduce the risk of frostbite. The nasal chamber might also conserve moisture during exhalation.

Fossil specimens from Laetoli in Tanzania and from Klasies River Mouth in South Africa indicate that anatomically modern *Homo sapiens* evolved sometime between about 200 and 100 kya in eastern or southern Africa. Molecular genetic data suggest that early *Homo sapiens* passed through a population bottleneck—that is, a period when they were rare creatures—before rapidly spreading throughout the Old World. They replaced indigenous hominin species in Eurasia, and then, as sea levels dropped during glacial periods, adventurous individuals went to sea in watercraft and populated Australia, the Americas, and oceanic islands.

Some of the extensive variation in bodily proportions, external features, and blood chemistry of modern peoples may reflect adjustments to biomes over geologically short time spans. However, molecular genetic studies show that genomic differences between even far-flung peoples are minuscule compared with variations within each local population. Accordingly, for modern *Homo sapiens*, race is a mere cultural construct with no biological basis [8, 9].

6.5. Language, Culture, and Life-Styles in the Pleistocene

The origin and development of human culture—articulate spoken language and symbolically mediated ideas, beliefs, and behaviour—are among the greatest unsolved puzzles in the study of human evolution. Such questions cannot be resolved by skeletal or archaeological data. Research on the behaviour and cognitive capabilities of apes, monkeys, and other animals and on cognitive development in human children provide some clues, but extrapolating this information back through time is tenuous at best. Complicating the scenario further, it may be that today's chimpanzees, bonobos, and other anthropoid primates have more sophisticated cognitive capabilities and behavioral skills than those of some early hominins, because they and their ancestors have had several million years to overcome many challenges and perhaps have become more advanced in the process.

6.6. Speech and Symbolic Intelligence

Speech has been inferred by some investigators on the basis of certain internal skull features, for example, in *H. habilis*, but jaw shape and additional traits suggest otherwise. Still other researchers claim that human speech was not even fully developed in early members of anatomically modern *Homo sapiens*, because of the simplicity of their tool kits and art before the Late Paleolithic [13].

It is impossible to assess linguistic competency by observing the insides of reassembled fossil craniums that are incomplete, battered, and distorted—and in any case the brains probably did not fit snugly against the walls of the braincase. The apparent cerebral expansion in *H. habilis* and *H. rudolfensis* may imply a general increase in cognitive abilities, manipulative skill, or other factors besides speech. Particularly unreliable are claims that the specific internal cranial impressions of a Broca cap is evidence of speech. Prominent Broca caps exist among some chimpanzees, yet no ape has uttered a word, despite laborious attempts to get them to speak [13, 14].

6.7. Lack of *Fossil* Evidence for Vocal Tract

A humanoid vocal tract is undetectable in fossils because it comprises only soft tissues and leaves no bony landmarks. Although versatile human speech is reasonably linked to a relatively spacious pharynx and mobile tongue, the absence of such features is not a compelling reason to deny some form of vocal language in ancestral hominins. It is argued that articulate human speech is impossible without a lowered voice box (larynx) and an expanded region above it. If this presumption were true, even Neanderthals would be inept vocally and probably also quite primitive cognitively as compared with Late Paleolithic *Homo sapiens* populations such as the Cro-Magnons. Gibbons and great apes do not speak, yet they have throat traits concomitant with speech, albeit to a lesser degree than humans'. The calls of gibbons are wonderfully varied in pitch and pattern, and, if such sounds were broken into discrete bits with consonants, they could emulate words. The same may be said for great apes. Orangutans, chimpanzees, and bonobos have sufficiently mobile lips and tongues; they simply lack neural circuitry for speech [15].

6.8. Multiple Intelligence Theory

Theory that different abilities are governed by distinct and separate forms of intelligence (multiple intelligences) is correct; much of tool-using behaviour and artistic ability would have to be based upon neurological structures fundamentally different from those that support verbal ability. Human children begin to use language before they become sophisticated tool users. Similarly, a form of speech might have preceded forms of tool behaviour that are symbolically mediated. Visual arts such as painting and sculpture are expressions of spatial intelligence, which is centered principally in areas of the brain different from those related to speech [16]. Therefore, one cannot expect the problem of language origins or language competence to be clarified by studying Paleolithic symbolism and imagery, despite the awesome array of cave art and polished bone, antler, ivory, stone, and shell artifacts associated with the period. Yet if the stunning proliferation and stylistic variability of tools, bodily ornaments, and artistic works during the Paleolithic do not point unequivocally to the specific use of speech, the presence of these symbolically mediated artifacts—among the earliest of which are shell beads found in Morocco and made about 82,000 years ago—does indicate that early humans were capable of complex conceptual and abstract thought [17].

6.9. System and Organization in Human Society

Historically, all human groups manifest rich symbolically mediated language, religion, and social, political, and economic systems, even in the absence of elaborate material culture. The demands on the social intelligence of peoples who live in environments with relatively few artifacts are similar to demands placed upon those who depend upon complex technological gadgets and shelters for comfort. Consequently, prehistoric *Homo sapiens* cannot be regarded as cognitively less capable than ourselves, and it is impossible to state which, hominin species were "fully human" as symbol users. As a case in point, meticulously documented language studies of captive bonobos and chimpanzees demonstrate that they have the capability to comprehend and use symbols in order to communicate with humans and with one another, but the use of this potential in the wild remains to be demonstrated. Perhaps the human capacity symbolically to represent feelings, situations, objects, and ideas developed before being commandeered by the several intelligences and before it became a boon to vocal communication [6, 17].

6.10. Feeding Habits

Archaeological evidence indicates that, like at least some of their Pliocene predecessors, the most recent hominins were probably omnivorous, though how much meat was in their diets and whether they obtained it by scavenging, hunting, or both are poorly documented until about 200–100 kya. Stone tools and cut marks on bones at archaeological sites attest to a long history of meat eating in the tribe Hominini, but this practice could have existed long before stone tools were invented. Like chimpanzees, bonobos, baboons, capuchins, and other primates, early Pliocene hominins may have killed and fragmented vertebrate prey with only their hands and jaws instead of tools. The extent to which our ancestors' hunting, scavenging, or other activities were communal and coordinated via symbolic communication has not been determined [18, 19].

6.11. Movement Patterns, Shelters, and Graves

There is no valid way to estimate group size and composition because there is little evidence of movement patterns, shelters, and graves until the Late Paleolithic. Archaeological traces of human-made shelters occur rarely from 60 kya, then become more common, particularly in regions with notable seasons of inclement weather. The first appearances and development of symbolically based spirituality are also highly elusive because they left no morphological or unarguable archaeological trace until the innovation of writing and ritual paraphernalia. Although some Neanderthals buried their dead, there is little evidence of mortuary ceremony in their graves. Graves of *Homo sapiens* from 40 kya sometimes contain grave goods [7, 19].

6.12. Learning from the Apes

Gorillas, chimpanzees, and bonobos are a rich resource for cultural anthropologists, biologists, and psychologists who speculate on the origins of human society. Gorillas appeal to theorists who stress male dominance and patriarchy. A characteristic gorilla group has one silverback (an older dominant male), one or more subordinate black-back males, adult females outnumbering males, and youngsters of various ages. The silverback is the hub of the cohesive group. Chimpanzee society is also dominated by males, which form a stable core of the group. Chimpanzees and bonobos live in larger groups numbering more than 100 individuals, though they forage, travel, and nest in much smaller bands that vary daily in number and composition. Among chimpanzees there is a top male, followed by several others whose ranks depend upon which other males are present. Bonobos have stronger affiliations between males and females than chimpanzees do, and the organizational hub of bonobo social groups is based on intimate relations among adult females, particularly mothers, which often retain strong bonds with their sons. Adult male bonobos are less strongly bonded with one another than chimpanzee males are. Because bonobos are more pacific and tolerant in social relationships and are highly sexual, they are popular with those who would model our heritage as free of "killer apes." However, observers of apes, Old World monkeys, and other mammals have documented incidents of aggression as well as concern for others in their subjects. Both tendencies are deeply rooted among the higher primates [7, 20].

The emergence of the human nuclear family has been a particularly knotty problem for Western evolutionary theorists. Like bonobos and chimpanzees, people probably are fundamentally promiscuous, though such mating behaviour is heavily proscribed by the cultures into which individuals are born and reside. Indeed, theorists who wish to construct models of the emergence of hominin societies on the basis of extant ape societies seldom tackle the overriding fact that humans utilize a wide variety of kinship, social, sexual, and political arrangements, all of which are maintained and expressed symbolically as well as practically. Researchers often fail to search for the cognitive basis of symbolic representation, manipulation, and invention in apes, citing instead forms of behaviour that appear to harbinger specific human conditions. It will take the efforts of several scientific disciplines and sophisticated technology, probably over many years, to discover the underlying nature of our mental faculties, their neurological basis, and their development over time. Apes can play important roles in this enterprise only if they are allowed to survive in their natural habitats and only if they are viewed as being on their own evolutionary paths and not merely as steps toward the human condition [19].

6.13. Why We Get Autism but Our Neanderthal Cousins Didn't

It's not what you've got but how you use it. The first maps of gene expression in two of our extinct cousins flag up important differences between the activity of their genes and our own. The results suggest that brain disorders like schizophrenia and autism may be unique to us.

The genomes of Neanderthals and Denisovans are relatively similar to our own, although we know that Neanderthals had some distinctive features, including a barrel-shaped chest and proportionally shorter limbs. Perhaps that's because these features were sculpted primarily by differences in the activity of genes during development rather than by differences in the genetic code.

Now Liran Carmel and his colleagues at the Hebrew University of Jerusalem in Israel have begun to investigate this possibility. They have created the first maps of gene activity in these extinct people, and compared them with modern humans.

Chris Stringer at the Natural History Museum in London hails the work as a remarkable achievement. "This pioneering work promises many further breakthroughs in understanding the biology of pre-modern humans," he says.

6.13.1. Autism Link

The activity of a gene typically decreases if it gets tagged with a chemical containing a methyl group – a process known as methylation.

We already know that methylated DNA bases decay in a distinctive manner, so by working through the Neanderthal and Denisovan genomes – and genomes of modern humans – Carmel and his colleagues could identify about 2000 genes that had been methylated differently in the three lineages.

Their results show that methylation in modern humans is particularly likely to affect genes with known links to health and neurological disorders – suggesting that the activity of these genes is different in our lineage to how it was in Neanderthals and Denisovans.

"We see associations with diseases like schizophrenia, autism and Alzheimer's," says Carmel. "Could it be that recent changes in the activity of genes in our brain also led to psychiatric disorders?"

He stresses that answering this question is difficult using the new data, because it is hard to predict how the differing pattern of gene activity would manifest itself in the brain.

Bernard Crespi and his group in 2002 at Simon Fraser University in Burnaby, Canada, studied the evolution of human neurodevelopment. He says the current thinking does indeed suggest that these disorders may be uniquely human. "Autism and schizophrenia are commonly viewed as disorders involving human-specific traits such as language and complex social cognition," he says. "As such, they are expected to be human-specific at least to a considerable degree" [21].

6.14. Short and Stout

The new information provides tantalising clues that might inform the search for more Denisovan fossils: although we have samples of their DNA, to date we know of just one Denisovan finger bone and two unusually large teeth. The new analysis shows that Neanderthals and Denisovans both had distinctive patterns of methylation in genes involved in limb development not seen in our species.

As far as Neanderthals go, these distinctive patterns might help to explain why their limbs were shorter and stouter than our own, says Carmel. It is possible, then, that Denisovans also had distinctive limbs – although Carmel says the pattern of methylation in the Denisovan DNA is sufficiently different from the pattern in Neanderthals to make it unclear exactly how the Denisovan limbs might have looked.

The study is not the first attempt to look at gene expression in prehistoric remains. In 2012, Alan Cooper at the University of Adelaide in Australia and Catherine Suter at the University of New South Wales in Sydney, Australia, looked at methylation patterns in 26,000-year-old bison DNA. Such methylation patterns can actually be inherited – a process known as epigenetic inheritance – and Cooper thinks this might have allowed animals to adapt to the rapidly changing climates of the ice age faster than traditional genetic inheritance could.

"We are trying to track the impacts of major climatic shifts on the epigenetic variation in vertebrate populations," says Cooper. "I'm convinced this is how many of the megafaunal groups, including humans, adapted to the major and repeated climate swings of the late Pleistocene" [22].

6.15. Ancient Warfare: Fighting for the Greater Good

War, what is it good for? A lot, it could turn out.

Lethal warfare drove the evolution of altruistic behaviour among ancient humans, claims a new study based on archaeological records and mathematical simulations.

If correct, the new model solves a long-standing puzzle in human evolution: how did our species transition from creatures interested in little more than passing down their own genes to societies of (generally) law-abiding (mostly) monogamists?

No one knows for sure when these changes happened, but climactic swings that occurred between approximately 10,000 to 150,000 years ago in the late Pleistocene period may have pushed once-isolated bands of hunter-gatherers into more frequent contact with one another, says Samuel Bowles, an evolutionary biologist at the Santa Fe Institute in New Mexico and the University of Siena, Italy, who led the study. "I think that's just a recipe for high-level conflict."

6.15.1. Tribes at War

By warfare, Bowles isn't talking about highly organised contests between nation-states and their armies. Rather, this period of warfare was probably characterised by ongoing skirmishes between neighbouring populations.

"We're talking about groups of men who got out in twos or threes or fives," he says. "They didn't have a chain of command and it's hard to see how they could force people to fight."

For this reason, altruistic intent on the part of each warrior is key. Each person would do better to stay home than to put their life on the line for their neighbours – yet they still went out and risked their lives, Bowles says.

To assess whether or not people with a random genetic predisposition to altruism could flourish via armed conflicts, Bowles culled archaeological and ethnographic data on the lethality of ancient warfare and plugged them into an evolutionary model of population change.

6.15.2. Cost of Clashes

In ancient graves excavated previously, Bowles found that up to 46 per cent of the skeletons from 15 different locations around the world showed signs of a violent death. More recently, war inflicted 30 per cent of deaths among the Ache, a hunter-gatherer population from Eastern Paraguay, 17 per cent among the Hiwi, who live in Venezuela and Colombia, while just 4 per cent among the Anbara in northern Australia.

On average, warfare caused 14 per cent of the total deaths in ancient and more recent hunter-gatherers populations.

The cost of losing an armed conflict as a group is high enough to balance out the individual risks of warfare, especially if a population is relatively inbred, Bowles' model concludes. Since evolution acts on genes, it makes more sense to make more sacrifices for a related neighbour than an unrelated one.

Since Bowles had no way of knowing how inbred Pleistocene populations were, he compared contemporary hunter-gatherers such as African pygmies and native Siberians. Individuals in these populations were closely related enough to justify going to war, he found.

6.16. INBREEDING

"There's no doubt that his is a controversial view," says Ruth Mace, an evolutionary anthropologist at University College London. Inbreeding between the victors and any surviving losers would dilute, not concentrate, altruistic genes, she says.

Bowles modelled this possibility in a previous paper and found that even with a measure of inbreeding, altruists still win out. However, he agrees that it would slow the evolution of altruism through warfare. "A much better way to spread the genes is to kill everybody," he says.

Mark van Vugt, a psychologist at the University of Kent at Canterbury, UK, notes that warriors could act in their own self-interest, not for the good of the group.

"Studies on the Amazonian Yąnomamö people show that these warriors do get a greater share of resources, they get more women, they sire more offspring," he says. "How do you explain that there are individual benefits for these warriors? There shouldn't be."

Still, van Vugt thinks Bowle's model is on the right track. Studies show that people divided into arbitrarily chosen groups – say heads and tails – behave altruistically to members of their group, but are more hostile toward non-members [23].

"Together we provide different pieces of the puzzle. If they fit together, they are starting to make sense," van Vugt says.

6.17. Stone Age DNA Shows Hunter-Gatherers Shunned Farming

How did farming take over the world? One theory is that farming was such an evidently good idea that it spread on its own. As hunter-gatherers encountered farmers, they were converted to an agricultural way of living. But a large-scale genetic analysis of Stone Age remains in Sweden suggests that this wasn't the case.

Instead, it seems like early farmers and hunter-gatherers had deep-rooted genetic differences. This suggests that European farmers were so successful that they displaced hunter-gatherers as they spread across the continent.

Pontus Skoglund of Uppsala University in Sweden and his colleagues sequenced the DNA from 11 early hunter-gatherers and farmers dating back to between 5000 and 7000 years ago. Four were associated with late Stone Age farming settlements; seven were identified as coastal hunter-gatherers [24].

DNA analysis showed that the farmers and hunter-gatherers descended from distinct genetic lineages. "It is quite clear that the two groups are very different," says Skoglund. Comparisons with the genes of modern populations revealed them to be more distinct that the genomes of modern Scandinavians and Italians.

Previous analyses of the isotopes in the bones of the 11 Stone Age individuals also showed the hugely different diet the two groups had. The hunter-gatherers relied primarily on seals and fish, while the farmers ate mostly land protein – presumably from the animals that they took care of [24].

6.18. Assimilations

"More interesting, here we find much lower genomic variation among the hunter-gatherers than among the farmers," says Anders Götherström of Stockholm University, one of the authors of the paper.

That means the hunter-gatherers were all very closely related to each other, which generally indicates a small population living in small groups.

Götherström thinks there are a few possible explanations. One is that after humans left Africa, some flourished by becoming farmers and exploiting the fertile lands they found in the Near East. Others ventured further north, where they continued to rely on the wild animals they killed and the plants and berries they collected from the landscape. These groups were split into small populations that survived in isolation between the ice sheets.

The Swedish data shows some mixing between the two groups, suggesting that as the farmers later swept north they interbred with the hunter-gatherers they encountered and assimilated them into their cultures. The exchange seems to have been unidirectional: there is very little if any evidence that farmers were assimilated into the hunter-gatherers.

The findings support previous studies of other early farmer and hunter-gatherer populations in Germany and elsewhere, says Joachim Burger of Johannes Gutenberg University of Mainz in Germany [24].

6.19. Neanderthals May Have Been Our Intellectual Equals

Enough of the cheap jibes: Neanderthals may have been just as clever as modern humans. Anthropologists have already demolished the idea that Neanderthals were dumb brutes, and now a review of the archaeological record suggests they were our equals.

Neanderthals were one of the most successful of all hominin species, occupying much of Europe and Asia. Their final demise about 40,000 years ago, shortly after *Homo sapiens* walked into their territory, is often put down to the superiority of our species.

It's time to lay that idea to rest, say Paola Villa at the University of Coloradoin Boulder and Wil Roebroeks at Leiden University in the Netherlands.

6.20. How to Speak Neanderthal: Perhaps We Do Already

We know our ancestors cosied up to our Neanderthal cousins, but did their liaisons leave an imprint on today's languages?

IF YOU find yourself stuttering your way through tourist French, spare a thought for the first modern humans. Travelling from Africa to Asia and Europe about 70,000 years ago, they would have encountered their evolutionary cousins, the Neanderthals, for the first time.

What did they say? In the past, many would have answered "not a lot" since Neanderthals weren't thought to have complex speech. But recent evidence suggests they probably had languages very similar to our own. Surprisingly, we may now have the means to glimpse those utterances in the words we speak today, with huge consequences for our understanding of language evolution.

The argument that Neanderthals spoke like us comes from many discoveries. Archaeological remains show that they had a sophisticated lifestyle, with human traits.

6.21. Just As Smart As You

For instance, there is evidence that *Homo sapiens* could use fire to chemically transform natural materials into glue 70,000 years ago, but Neanderthals were performing similarly complex chemical syntheses at least 200,000 years ago.

And although 70,000-year-old engraved ochre from South Africa is seen as evidence that our species had developed sophisticated symbolism and perhaps even language, similar artefacts have been found at 50,000-year-old Neanderthal sites in Spain.

What's more, Neanderthals might have been able to talk. Late last year we learned that our extinct cousins had a hyoid, a small bone in the neck that plays a big role in speech, very like ours.

Evidence has even emerged that *Homo sapiens* may have learned some skills by copying Neanderthals. Yet despite all of this evidence, the idea that Neanderthals were our inferiors still persists.

6.22. Fossilized Behavior

Archaeologists are reluctant to accept evidence of advanced behaviour if it is attributed to extinct hominins, says Roebroeks. This prevailing attitude influences our ideas about the causes of the Neanderthal extinction.

Villa and Roebroeks say this is not the case for other transitions in the archaeological record, such as the arrival of farming in Europe and the subsequent disappearance of Europe's hunter-gatherers. Our ideas about the disappearance of European hunter-gatherers do not assume that they were less intelligent than farmers.

Villa and Roebroek are right to apply the same standards to Neanderthal and human archaeological sites, says Francesco d'Errico at the University of Bordeaux in France. "This makes their argument in favour of equal cognitive abilities more robust than previous attempts."

However it may be that the archaeological evidence simply isn't good enough to identify the differences between Neanderthal and modern human behaviour, says Shannon McPherron at the Max Planck Institute for Evolutionary Anthropology in Leipzig, Germany [25]. "*I believe there were significant differences between Neanderthals and moderns and that this is why we have the replacement of one by the other*," he says. "*But it is still a challenge to see this in the archaeological record and demonstrate it properly*."

REFERENCES

[1] Mead S, Whitfield J, Poulter M, et al. (2009). A Novel Protective Prion Protein Variant that Colocalizes with Kuru Exposure. *N Engl J Med* 36, 2056–2065.

[2] Byars SG, Ewbank D, Govindaraju DR, Stearns SC (2009). Natural selection in a contemporary human population. *Pr Natl Acad Sc* 107 (suppl_1): 1787–1792.

[3] Cochran G, Harpending H (2009). *The 10,000 Year Explosion*. Basic Books N.Y.

[4] Diamond J (1992). *The Third Chimpanzee*. Harper Perennial. pp. 47–57.

[5] Diamond J (1999). *Guns, Germs, and Steel: The Fate of Human Societies*. W. W. Norton. p. 39.

[6] Leakey R, Lewin R (1992). *Origins Reconsidered: In Search of What Makes us Human*. Little, Brown and Company.

[7] Shreeve J (1996). *The Neanderthal Enigma: Solving the Mystery of Modern Human Origins*. Viking.

[8] Stringer C, McKie R (1996). *African Exodus: The Origins of Modern Humanity*. Jonathan Cape.

[9] Walker A, Shipman P (1996). *The Wisdom of the Bones: In Search of Human Origins*. Weidenfeld and Nicholson.

[10] Wells S (2004). *The Journey of Man: A Genetic Odyssey* (in English). New York, NY: Random House Trade Paperbacks.

[11] Wade N (2006). *Before the Dawn: Recovering the Lost History of our Ancestors*. Penguin Press.

[12] Tattersall I (2008). *The Fossil Trail: How We Know What We Think We Know About Human Evolution*. Oxford University Press.

[13] Enard W, Przeworski M, Fisher SE, Lai CSL et al. (2002). Molecular Evolution of FOXP2, a gene involved in speech and language. *Nature* 418, 869-872.

[14] Flinn MV, Geary DC, Ward CV (2005). Ecological dominance, social competition, and coalitionary arms races: Why humans evolved extraordinary intelligence. *Evol Hum Behav*, 26, 10-46.

[15] Johanson D, Wong K (2009). *Lucy's Legacy: The Quest for Human Origins*. Three Rivers Press.

[16] Jones S, Martin RD, Pilbeam DR (Editors) (1994). *The Cambridge Encyclopedia of Human evolution*. Cambridge University Press.

[17] Lewin R (1987). *Bones of Contention: Controversies in the Search for Human Origins*. Penguin Books.

[18] Morwood M, van Oosterzee P (2007). *A New Human: The Startling Discovery and Strange Story of the 'Hobbits' of Flores, Indonesia*. Smithsonian Books.

[19] Roberts A (2009). *The Incredible Human Journey: The Story of how we Colonised the Planet*. Bloomsbury.

[20] Klein RG (1999). The Human Career: Human Biological and *Cultural Origins*, 2nd ed.

[21] Crespi BP, Stead P, Elliot M (2010). Comparative genomics of autism and schizophrenia. Proc Nat Acad Sc USA 107, 1736-1741.

[22] Gokhman D, Lavi E, Prüfer K et al. (2014). Reconstructing the DNA Methylation Maps of the Neandertal and the Denisovan. Science 344, 523-527.

[23] Bowles S (2009). Did Warfare Among Ancestral Hunter-Gatherers Affect the Evolution of Human Social Behaviors? *Science* 324, 1293-1298.

[24] Skoglund P, Malmström H, Omrak A, et al. (2014). Genomic Diversity and Admixture Differs for Stone-Age Scandinavian Foragers and Farmers. *Science* 344, 747-750.

[25] Villa P, Roebroeks W (2014). *Neandertal Demise: An Archaeological Analysis of the Modern Human Superiority Complex*. DOI: 10.1371/ journal.pone.0096424/

Chapter VII

IDA AS 'MISSING-LINK' OF ANTHROPOID EVOLUTION

ABSTRACT

Researchers have long searched for the earliest anthropoids, advanced primates that were the ancestors of humans, apes, and monkeys. Until recently, most scientists thought anthropoids arose in Africa, where the oldest widely accepted members of the group lived as early as 37 million years ago in the Fayum province of Egypt. But recent evidence has pointed to an Asian origin. Fossils from that continent suggest that the first anthropoids could have been either *Eosimias*, a genus of tiny primates that lived 4 million to 45 million years ago, or a younger group called amphipithecids. Discovery of Ida in an oil shale pit in Germany suggested to a handful of paleontologists that a third group of primates, called adapids, might have given rise to the first primitive anthropoids instead, this time from Europe – although few experts in primate origins agreed.

7.1. DISCOVERY OF IDA

On May 20 of last year, at a remarkable press conference in New York, a group of researchers announced—with much ballyhoo—that they'd found a 47-million-year-old primate fossil named *Darwinius masillae* (in honor of Charles Darwin nicknamed "Ida" after his daughter). Ida's skeletal remains are remarkably complete, putting her in a small, elite group of well-documented fossil primates from the Eocene (55 to 34 million years ago) that also includes her North American cousin, *Notharctus*. Uniquely for primate fossils this old, Ida's stomach contents and a few aspects of her soft anatomy are preserved. Despite the hype, Ida looks nothing like a transitional "apeman," instead looking quite like a modern lemur. Ida, the finest fossil primate in existence, was touted loudly as the missing link between the two major branches of primates, the Haplorhini (anthropoids [apes and monkeys] and tarsiers), and the Strepsirrhini (lemurs and lorises; see figure below) (Figure 1).

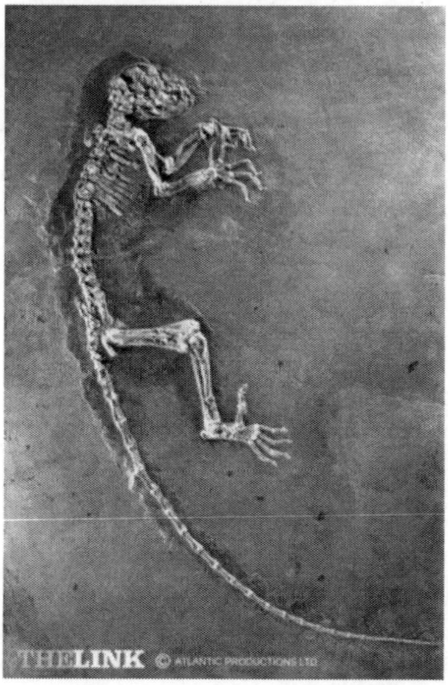

Figure 1. Ida (*Darwinius masillae*) (Adapted from Franzen et al. 2009, *PLoS ONE* 4, e5723).

Concurrent with the press conference was a History Channel documentary and a book about Ida, Colin Tudge's *The Link*, that proclaimed, with much heavy breathing, that Ida was, as one of the earliest primate ancestors of our own species, an earthshaking discovery (see my review of the book here). The Ida fossil, which was found in the Messel Pit on the outskirts of Frankfurt, was revealed to the public in what amounted to the greatest publicity coup in modern science. The mayor of New York, Michael Bloomberg, appeared alongside the fossil, wearing a T-shirt carrying the TV tie-in logo, "The link." A book about Ida was already coming off the presses. Ida was an immediate media sensation. The fossil received blanket coverage around the world and newspapers hailed her as the "missing link" between humans and animals. The

Guardian even gave away free wallcharts of "humanity's long lost ancestor" [1].

7.2. New Discoveries to Refute the Claims

The controversy erupted after Seiffert's team unearthed the fossilised remnants of a similar, but much younger primate in northern Egypt. Analysis of the 37million-year-old lemur-like fossil showed it was a close relative of Ida and had several dental features that are commonly seen in apes and humans (Figure 2).

Now, the newest find tips the scales back in favor of the Asian origin of anthropoids. In 2005, Beard discovered a 37-million-year-old jaw fragment in the badlands of central Myanmar. The jaw belonged to an amphipithecid, which Beard and paleontologist Jean-Jacques Jaeger of the University of Poitiers in France have named *Ganlea megacanina* in the *Proceedings of the Royal Society B*. Although the fossil is not nearly as dazzling as a complete skeleton of Ida, Beard knew immediately that the jaw was something special. It has a huge canine tooth, which was used to pry open the hard shell of tough tropical fruits to eat the seeds, according to wear markings (Figure 2). That's an unusual feeding adaptation found in modern saki monkeys in the Amazon Basin of South America [1, 2].

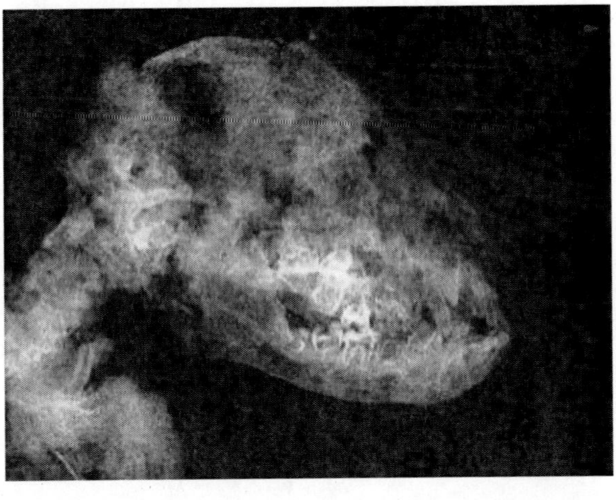

Figure 2. Close up of skull indicating Ida's youth (less than a year old). Her milk teeth are visible and she shows teeth visible in keeping with plant and fruit eaters (Adapted from Franzen et al. 2009, *PLoS ONE* 4, e5723).

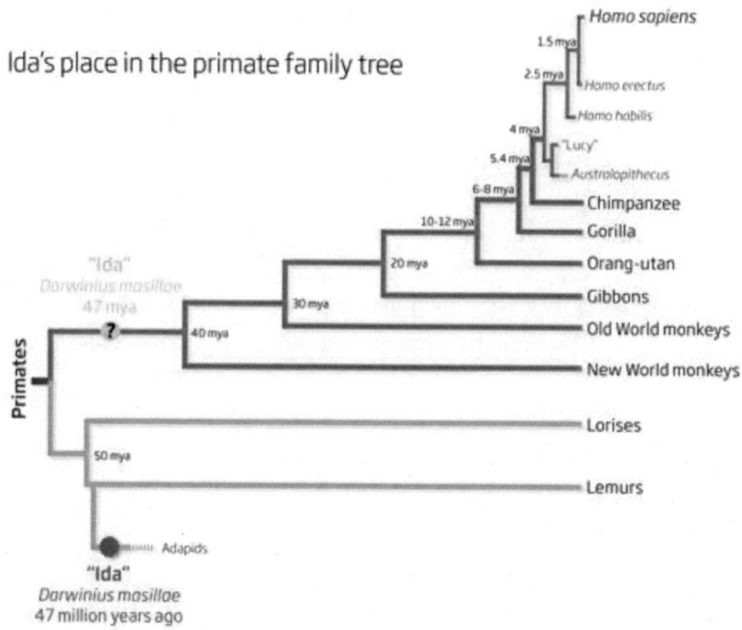

Figure 3. Primate family tree (Beard 2009).

7.3. Asia, the Cradle of Anthropoid Evolution

This monkeylike behavior, as well as anatomical features in other fossils of amphipithecids from Asia, adds new evidence to the view that amphipithecids were early anthropoids and, hence, that anthropoids arose in Asia, says Beard. Others agree: These fossils "affirm the importance of Asia as a hotbed of early anthropoid evolution." Not so fast, says paleoanthropologist Russell Ciochon of the University of Iowa in Iowa City. Although he agrees that anthropoids arose in Asia, he would like to see more evidence than just a fragmentary jaw. The new fossil, he says, does "nothing to provide resolution to the anthropoid origins debate." Ross counters that the new fossil is indeed important. It "provides important independent confirmation that Asian amphipithecids are closely related to [the most primitive] anthropoids [3].

7.4. IDA IS NOT A 'MISSING-LINK'

University of Oslo paleontologist Jørn Hurum, co-author of the *PLoS* report detailing the fossil, claimed, "This fossil is part of our evolution." Although Ida appears very lemur-like, Hurum considers her transitional to haplorhines, a broad grouping of primates thought to exclude lemurs but include monkeys and apes. However, Chris Beard of the Carnegie Museum of Natural History wrote that "Ida is not a 'missing link' – at least not between anthropoids and more primitive primates" [1]. And Duke University paleontologist Richard Kay stated, "There is no phylogenetic analysis to support the claims" that Ida is a missing link [1, 3] (Figures 3, 4).

Not only does Ida not have transitional features that a missing link should have, but she was found in a rock layer that is too young for her to even be a candidate for a "missing link" status. Harvard-trained paleontologist Kurt Wise told the *Baptist Press* that, according to evolutionary timescales, "Ida is much younger than both good fossils of lemurs and good fossils of monkeys." Seiffert's team fed information from the new fossil and 117 living and extinct primates into a computer model to find out where the new species sat in the tree of life. Writing in Nature, Seiffert explains that while the new fossil, named Afradapis, is related to Ida, both emerged along the evolutionary path that led to lemurs and lorises. Their anatomical similarities with later primates evolved independently from those seen in monkeys, humans and apes [1, 2].

"They are trying to explain all of the traits we see in Darwinius in terms of parallel evolution," said Hurum. Parallel evolution is when two groups of animals evolve similar features without being related to one another. According to Philip Gingerich, a leading paleontologist at Princeton University who worked on Ida, both fossils were almost certainly part of the lineage that led to monkeys, apes and humans. It was "puzzling" to see Seiffert's team claim they were related to a group that became lemurs and lorises "with which it shares no resemblance." Further work by Seiffert's team appears to add insult to injury. According to their study, neither Ida nor Afradapis have any living descendants, meaning they became extinct at the end of a sidebranch of the evolutionary tree (Figures 3, 4).

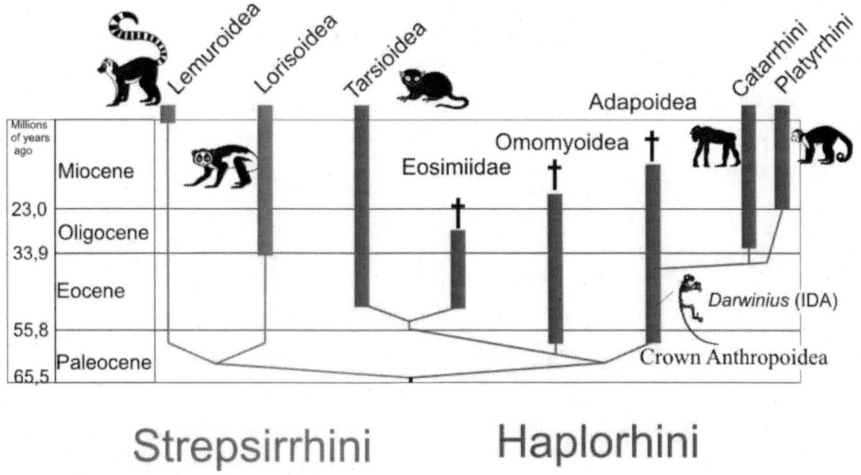

Figure 4. Position of *Darwinius* in the primate phylogeny (Adapted from Gingerich et al. 2010, *J. Hum. Evol.* 59, 574-579).

7.5. Why Ida Fossil Is Not the Missing Link

The description of Ida was eventually published in a paper by Franzen et al. in the journal *PLoS ONE*, (Figure 2). The reaction of both bloggers and scientists was very critical, with many pointing out that Ida didn't look like a missing link at all, but may have been only an adapiform primate, a representative of a lineage *within* the Strepsirrhini that went extinct without leaving descendants. The phylogenetic analysis of the *PLoS* paper, they claimed, was cursory and incomplete. A later paper in *Nature* by Seiffert et al. cast further doubt on Ida's status, suggesting that the features that led Ida's discoverers to lump her with the haplorhines were convergent—that they had evolved independently in haplorhines and adapiforms and thus could not serve as evidence for Ida's anthropoid ancestry [1, 4-6].

In an article now available online in the *Journal of Human Evolution*, four scientists present evidence that the 47-million-year-old *Darwinius masillae* is not a haplorhine primate like humans, apes and monkeys, as the 2009 research claimed. They also note that the article on *Darwinius* published last year in the journal *PLoS ONE* ignores two decades of published research showing that similar fossils are actually strepsirrhines, the primate group that includes lemurs and lorises [4-6]. Summarizing all the evidence in a cladistic framework (combining, as Williams et al. say, "hundreds of dental, skull, limb-bone, embryological, physiological and molecular characters of living and fossil primates"), they come up with two conclusions (Figure 4):

1. First, the Eocene adapids (of which Ida was one) were not, as Franzen et al. implied, more closely related to anthropoid primates than to lemurs and lorises. Rather, the adapids were stem strepsirrhines that lived well after that group diverged from the haplorhines (our own ancestors) [1, 4-6].
2. Second, detailed study of Ida's mandibles, teeth, orbit, leg bones and other traits show that she possesses no clear *synapomorphies* (shared derived traits) with haplorhines, but rather is clearly an adapiform. All the evidence, then (more will soon be published in *Proceedings of the National Academy of Sciences*) puts Ida far from the haplorhines and anthropoids, and firmly in the now-extinct adapids. Franzen's failure to recognize this apparently rests partly on their unaccountable failure to use *any* fossil haplorhines in their original phylogenetic analysis.

The well-preserved fossil (95 percent complete, including fossilized fur and more) is about the size of a raccoon and includes a long tail. It resembles the skeleton of a lemur (a small, tailed, tree-climbing primate). The fossil does not resemble a human skeleton.

- The fossil was found in two parts by amateur fossil hunters in 1983. It eventually made its way through fossil dealers to the research team.
- Ida has opposable thumbs, which the ABC News article states are "similar to humans' and unlike those found on other modern mammals" (i.e., implying that opposable thumbs are evidence of evolution). Yet lemurs today have opposable thumbs (like all primates). Likewise, Ida has nails, as do other primates. And the talus bone is described as "the same shape as in humans," despite the fact that there are other differences in the ankle structure (Figure 5).
- Unlike today's lemurs (as far as scientists know), Ida lacks the "grooming claw" and a "toothcomb" (a fused row of teeth) In fact, its teeth are more similar to a monkey's. These are minor differences easily explained by variation within a kind (Figure 5).

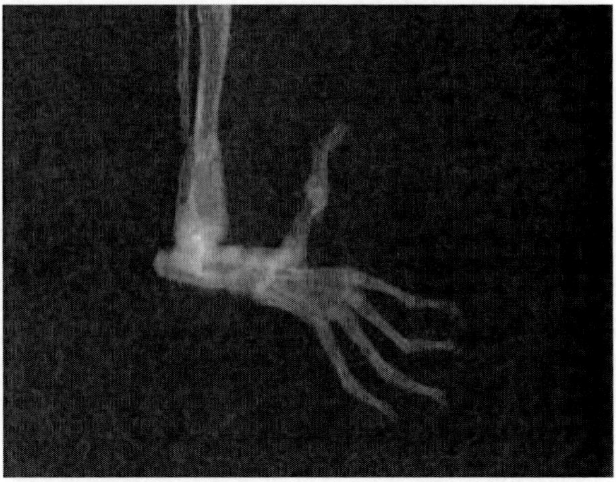

Figure 5. Close up showing the lack of grooming claw familiar with other lemur primates (Adapted from Franzen et al. 2009, *PLoS ONE* 4, e5723).

7.6. Evidence Against Ida As 'Missing-Link'

"Many lines of evidence indicate that *Darwinius* has nothing at all to do with human evolution," says Chris Kirk, associate professor of anthropology at The University of Texas at Austin. "Every year, scientists describe new fossils that contribute to our understanding of primate evolution. What's amazing about *Darwinius* is, despite the fact that it's nearly complete, it tells us very little that we didn't already know from fossils of closely related species." That literature centers on the evolution of primates, which include haplorhines (apes, monkeys, humans, tarsiers) and strepsirrhines (lemurs, lorises). The two groups split from each other nearly 70 million years ago [1, 7, 8].

The fossil group to which *Darwinius* belongs – the adapiforms – have been known since the early 1800s and includes dozens of primate species represented by thousands of fossils recovered in North America, Europe, Asia and Africa. Some adapiforms, like North American Notharctus, are known from nearly complete skeletons like that of *Darwinius*. Most analyses of primate evolution over the past two decades have concluded that adapiforms are strepsirrhines, and not direct ancestors of modern humans. The most recent such analysis, published last year in the journal Nature, concluded that *Darwinius* is an early strepsirrhine and a close relative of the 39-million-year-old primate Mahgarita stevensi from West Texas. Nevertheless, the scientists who last year formally described *Darwinius* concluded that it was an early haplorhine, and even suggested that *Darwinius* and other adapiform fossils "could represent a stem group from which later anthropoid primates evolved." For example, they note that *Darwinius* has a short snout and a deep jaw – two features that are found in monkeys, apes, and humans. However, Kirk, Williams and their colleagues point out that short snouts and deep jaws are known to have evolved multiple times among primates, including several times within the lemur/loris lineage. They further argue that *Darwinius* lacks most of the key anatomical features that could demonstrate a close evolutionary relationship with living haplorhines (apes, monkeys, humans, and tarsiers). For instance, haplorhines have a middle ear with two chambers and a plate of bone that shields the eyes from the chewing muscles. "There is no evidence that *Darwinius* shared these features with living haplorhines [9, 10].

7.7. The Creationist Interpretation

The principles that inform creationists about Ida are some of the same that allow creationists to interpret fossil after fossil hailed as "transitional forms":

1. Nothing about this fossil suggests it is anything other than an extinct, lemur-like creature. Its appearance is far from chimpanzee, let alone "apeman" or human.
2. A fossil can never show evolution. Fossils are unchanging records of dead organisms. Evolution is an alleged process of change in live organisms. Fossils show "evolution" only if one presupposes evolution, then uses that presupposed belief to *interpret* the fossil.
3. Similarities can never show evolution. If two organisms have similar structures, the only thing it *proves* is that the two have similar structures. One must presuppose evolution to say that the similarities are due to evolution rather than design. Furthermore, when it comes to "transitional forms," the slightest similarities often receive great attention while major differences are ignored.
4. The remarkable preservation is a hallmark of rapid burial. Team member Jørn Hurum of the University of Oslo said, "This fossil is so complete. Everything's there. It's unheard of in the primate record at all. You have to get to human burial to see something that's this complete." Even the contents of Ida's stomach were preserved. While the researchers believe Ida sunk to the bottom of a lake and was

buried, this preservation is more consistent with a catastrophic flood. Yet Ida was found with "hundreds of well-preserved specimens [5]."

5. If evolution were true, there would be ***real*** transitional forms. Instead, the best "missing links" evolutionists can come up with are strikingly similar to organisms we see today, usually with the exception of minor, controversial, and inferred anatomical differences.
6. Evolutionists only open up about the lack of fossil missing links once a new one is found. Sky News reports, "Researchers say proof of this transitional species finally confirms Charles Darwin's theory of evolution," while Attenborough commented that the missing link "is no longer missing." So are they admitting the evidence was missing until now (supposedly) [4, 11-13]?

REFERENCES

[1] Beard C (2009). Why Ida fossil is not the missing link. *New Scientist* 2710, 18-19.

[2] Seiffert ER, Perry JMG, Simons EL, Boyer DM (2009). Convergent evolution of anthropoid in Eocene adapiform primates. *Nature* 461:118-1122.

[3] Gibbons A (2009). Revolutionary. Fossil Fails to Dazzle Paleontologists. *Science NOW Daily News.*

[4] Franzen JL, Gingerich PD, Habersetzer J et al. (2009). Complete primate skeleton from the middle Eocene of Messelin Germany: morphology and paleobiology. *PLoS ONE* 4, e5723.

[5] Gingerich PD, Frazen, JL, Habersetzer J, Hurum JH, Smith BH (2010). *Darwinius marsillae* is a Haplorhine- Reply to Williams et al. (2010). *J. Hum. Evol.* 59, 574-579.

[6] Williams BA, Kay RF, Kirk EC, Ross CF (2010). *Darwinius masillae* is astrepsirrhine—a Reply to Franzen et al. (2009). *J. Hum Evol.*

[7] Randerson J (2009a). Is fossil Ida a missing link in evolution? *The Guardian.*

[8] Randerson J (2009b). Fossil Ida: extraordinary find is 'missing link' in human evolution. *The Guardian.*

[9] Roach E (2009). Experts: Fossil find exciting but lacks significance. *Baptist Press.*

[10] Thomas B (2009). The Ida Fossil: A Clever Campaign for a Lackluster "Link." *Acts & Facts.* 38 (7): 17.

[11] Potter N (2009). Primate Fossil Could Be Key Link in Evolution. *ABC News*.
[12] Watts A (2009). Scientists Unveil Missing Link in Evolution. *Sky News Online*.
[13] Fossils from the Messel site. *The Guardian*, n.d.

APPENDIX I

HOW DID HUMANS EVOLVE?

About six million years ago in Africa, the chimpanzee lineage and our own split. What happened to us after that split? The hominid lineage did not march in a straight line to *Homo sapiens*. Instead, the early hominid lineage gave rise to many other (now extinct) hominids [1]. Examining the fossils, the artifacts, and even the DNA of these relatives has helped us understand how this complex hominid tree evolved, and how modern humans came to exist (Figure 1, Table 1).

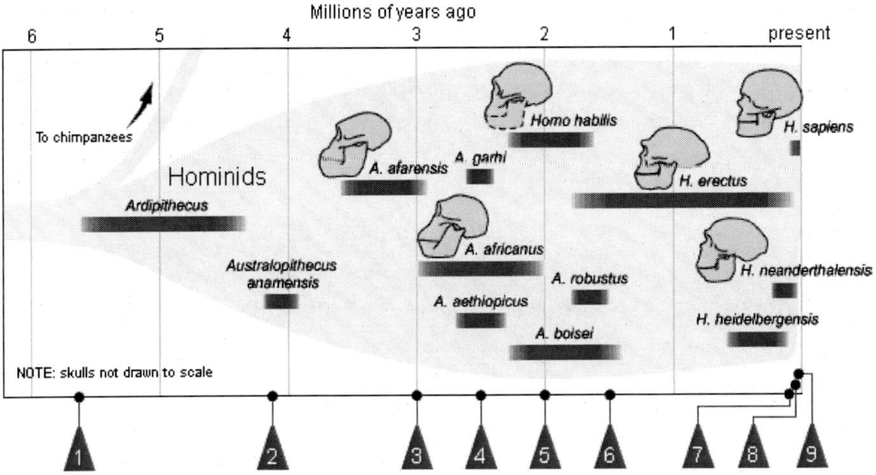

Figure 1. Time line of Human evolution (Adapted from Gibbons 2002, *Science* 295, 1214–1219).

Here are some of the important events in human history, with approximate dates, which reflect the evidence currently available:

1. **Before 5 mya:** In Africa, our ancestral lineage and the chimpanzee lineage split.
2. **Before 4 mya:** The hominid *Australopithecus anamensis* walked around what is now Kenya on its hind legs.
3. **>3 mya:** *Australopithecus afarensis* ("Lucy") lived in Africa.
4. **2.5 mya:** Some hominids made tools by chipping stones to form a cutting edge. There were perhaps four or more species of hominid living in Africa.
5. **2 mya:** The first members of the Homo clade, with their relatively large brains, lived in Africa.
6. **1.5 mya:** Hand axes were used. Also, hominids had spread out of Africa and into much of Asia and Europe. These hominids included the ancestors of Neanderthals (*Homo neanderthalensis*) in Europe and *Homo erectus* in Asia.
7. **100,000 years ago:** Human brains reached more or less the current range of sizes. Early *Homo sapiens* lived in Africa. At the same time, *Homo neanderthalensis* and *Homo erectus* lived in other parts of the Old World.
8. **50,000 years ago:** Human cultures produced cave paintings and body adornment, and constructed elaborate burials. Also, some groups of modern humans extended their range beyond Africa.
9. **25,000 years ago:** Other *Homo* species had gone extinct, leaving only modern humans, *Homo sapiens*, spread throughout the Old World .

REFERENCES

[1] Gibbons A (2002). In search of the first hominids. *Science* 295, 1214–1219.

APPENDIX II

Table 2. Evolution of Primates to Humans Primates

Date	Event
85–65 Ma	 *Plesiadapis*  *Carpolestes simpsoni* A group of small, nocturnal and arboreal, insect-eating mammals called the Euarchonta begins a speciation that will lead to the primate, treeshrew and flying lemur orders. The Primatomorpha is a subdivision of Euarchonta that includes the primates and the proto-primate Plesiadapiformes. One of the early proto-primates is *Plesiadapis*. *Plesiadapis* still had claws and the eyes located on each side of the head. Because of this they were faster on the ground than on the top of the trees, but they began to spend long times on lower branches of trees, feeding on fruits and leaves. The Plesiadapiformes very likely contain the species which is the ancestor of all primates [1]. One of the last Plesiadapiformes is *Carpolestes simpsoni*. It had grasping digits but no forward-facing eyes.

Table 2. (Continued)

Date	Event
63 Ma	Primates diverge into suborders Strepsirrhini (wet-nosed primates) and Haplorrhini (dry-nosed primates). Strepsirrhini contain most of the prosimians; modern examples include the lemurs and lorises. The haplorrhines include the three living groups: prosimian tarsiers, simian monkeys, and apes. One of the earliest haplorrhines is *Teilhardina asiatica*, a mouse-sized, diurnal creature with small eyes. The Haplorrhini metabolism lost the ability to make its own Vitamin C. This means that it and all its descendants had to include fruit in its diet, where Vitamin C could be obtained externally.
30 Ma	 *Aegyptopithecus* Haplorrhini splits into infraorders Platyrrhini and Catarrhini. Platyrrhines, New World monkeys, have prehensile tails and males are color blind. They may have migrated to South America on a raft of vegetation across the relatively narrow Atlantic ocean (approx. 700 km). Catarrhines mostly stayed in Africa as the two continents drifted apart. Possible early ancestors of catarrhines include *Aegyptopithecus* and *Saadanius*.
25 Ma	 *Proconsul* Catarrhini splits into 2 superfamilies, Old World monkeys (Cercopithecoidea) and apes (Hominoidea). Our trichromatic color vision had its genetic origins in this period. *Proconsul* was an early genus of catarrhine primates. They had a mixture of Old World monkey and ape characteristics. *Proconsul*'s monkey-like features include thin tooth enamel, a light build with a narrow chest and short forelimbs, and an arboreal quadrupedal lifestyle. Its ape-like features are its lack of a tail, ape-like elbows, and a slightly larger brain relative to body size. *Proconsul africanus* is a possible ancestor of both great and lesser apes, including humans.

HOMINIDAE

Date	Event
15 Ma	Hominidae (great apes) speciate from the ancestors of the gibbon (lesser apes).
13 Ma	Homininae ancestors speciate from the ancestors of the orangutan [2]. *Pierolapithecus catalaunicus* is believed to be a common ancestor of humans and the great apes or at least a species that brings us closer to a common ancestor than any previous fossil discovery. It had special adaptations for tree climbing, just as humans and other great apes do: a wide, flat rib cage, a stiff lower spine, flexible wrists, and shoulder blades that lie along its back.
10 Ma	The lineage currently represented by humans and the Pan genus (chimpanzees and bonobos) speciates from the ancestors of the gorillas.
7 Ma	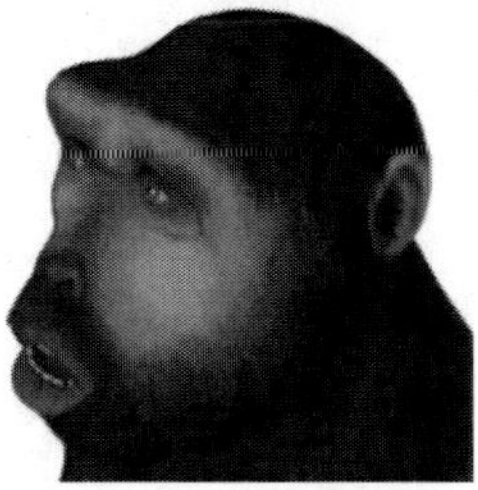 *Sahelanthropus tchadensis* Hominina speciate from the ancestors of the chimpanzees. Both chimpanzees and humans have a larynx that repositions during the first two years of life to a spot between the pharynx and the lungs, indicating that the common ancestors have this feature, a precondition for vocalized speech in humans. The latest common ancestor lived around the time of *Sahelanthropus tchadensis*, ca. 7 Ma [4]; *S. tchadensis* is sometimes claimed to be the last common ancestor of humans and chimpanzees, but there is no way to establish this with any certainty. The earliest known representative from the ancestral human line post-dating the separation with the chimpanzee lines is *Orrorin tugenensis* (Millennium Man, Kenya; ca. 6 Ma).
4.4 Ma	*Ardipithecus* is a very early hominin genus (subfamily Homininae). Two species are described in the literature: *A. ramidus*, which lived about 4.4 million years ago [3] during the early Pliocene, and *A. kadabba*, dated to approximately 5.6 million years ago (White et al. 2009) (late Miocene). *A. ramidus* had a small brain, measuring between 300 and 350 cm^3. This is about the same size as modern bonobo and female common chimpanzee brain, but much smaller than the brain of australopithecines like Lucy (~400 to 550 cm^3) and slightly over a fifth the size of the modern *Homo sapiens* brain. Ardipithecus was arboreal, meaning it lived largely in the forest where it competed with other forest animals for food, including the contemporary ancestor for the chimpanzees.

Table. (Continued)

Date	Event
	Ardipithecus was probably bipedal as evidenced by its bowl shaped pelvis, the angle of its foramen magnum and its thinner wrist bones, though its feet were still adapted for grasping rather than walking for long distances.
3.6 Ma	 *Australopithecus afarensis* Some *Australopithecus afarensis* left human-like footprints on volcanic ash in Laetoli, Kenya (Northern Tanzania) which provides strong evidence of full-time bipedalism. *Australopithecus afarensis* lived between 3.9 and 2.9 million years ago. It is thought that *A. afarensis* was ancestral to both the genus *Australopithecus* and the genus *Homo*. Compared to the modern and extinct great apes, *A. afarensis* has reduced canines and molars, although they are still relatively larger than in modern humans. *A. afarensis* also has a relatively small brain size (~380–430 cm^3) and a prognathic (i.e., projecting anteriorly) face. Australopithecines have been found in savannah environments and probably increased its diet to include meat from scavenging opportunities. An analysis of *Australopithecus africanus* lower vertebrae suggests that females had changes to support bipedalism even while pregnant.
3.5 Ma	*Kenyanthropus platyops*, a possible ancestor of *Homo*, emerges from the *Australopithecus* genus.
3 Ma	The bipedal australopithecines (a genus of the Hominina subtribe) evolve in the savannas of Africa being hunted by *Dinofelis*. Loss of body hair takes place in the period 3-2 Ma, in parallel with the development of full bipedalism.

HOMO

Date	Event
2.5 Ma	 *Homo habilis* Appearance of *Homo*. *Homo habilis* is thought to be the ancestor of the lankier and more sophisticated *Homo ergaster*. Lived side by side with *Homo erectus* until at least 1.44 Ma, making it highly unlikely that *Homo erectus* directly evolved out of *Homo habilis*. First stone tools, beginning of the Lower Paleolithic.
1.8 Ma	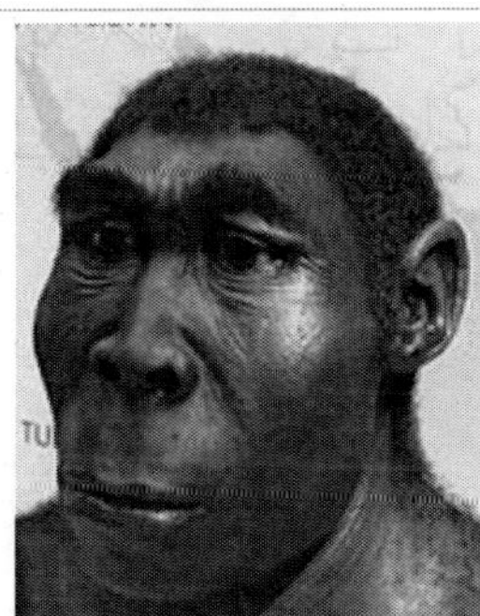 A reconstruction of *Homo erectus*. *Homo erectus* evolves in Africa. *Homo erectus* would bear a striking resemblance to modern humans, but had a brain about 74 percent of the size of modern man. Its forehead is less sloping than that of *Homo habilis* and the teeth are smaller. Other hominid designations such as *Homo georgicus*, *Homo ergaster*, *Homo pekinensis*, *Homo heidelbergensis* are often put under the umbrella species name of *Homo erectus*. Starting with *Homo georgicus* found in what is now the Republic of Georgia dated at 1.8 Ma, the pelvis and backbone grew more human-like and gave *H. georgicus* the ability to cover very long distances in order to follow herds of other animals.

Table. (Continued)

Date	Event
	This is the oldest fossil of a hominid found outside of Africa. Control of fire by early humans is achieved 1.5 Ma by *Homo ergaster. Homo ergaster* reaches a height of around 1.9 metres (6.2 ft). Evolution of dark skin, which is linked to the loss of body hair in human ancestors, is complete by 1.2 Ma. *Homo pekinensis* first appears in Asia around 700 Ka but according to the theory of a recent African origin of modern humans, they could not be human ancestors, but rather, were just a cousin offshoot species from *Homo ergaster. Homo heidelbergensis* was a very large hominid that had a more advanced complement of cutting tools and may have hunted big game such as horses
1.2 Ma	*Homo antecessor* may be a common ancestor of humans and Neanderthals [4, 5]. At present estimate, humans have approximately 20,000–25,000 genes and share 99% of their DNA with the now extinct Neanderthal [6] and 95-99% of their DNA with their closest living evolutionary relative, the chimpanzees [7]. The human variant of the FOXP2 gene (linked to the control of speech) has been found to be identical in Neanderthals [8]. It can therefore be deduced that *Homo antecessor* would also have had the human FOXP2 gene.
600 ka	 A reconstruction of *Homo heidelbergensis* Three 1.5 m (5 ft) tall *Homo heidelbergensis* left footprints in powdery volcanic ash solidified in Italy. *Homo heidelbergensis* may be a common ancestor of humans and Neanderthals [9]. It is morphologically very similar to *Homo erectus* but *Homo heidelbergensis* had a larger brain-case, about 93% the size of that of *Homo sapiens*. The holotype of the species was tall, 1.8 m (6 ft) and more muscular than modern humans. Beginning of the Middle Paleolithic.
338 ka	Y-chromosomal Adam lived in Africa approximately 338,000 years ago, according to a recent study [10]. He is the most recent common ancestor from whom all male human Y chromosomes are descended.

Date	Event
200 ka	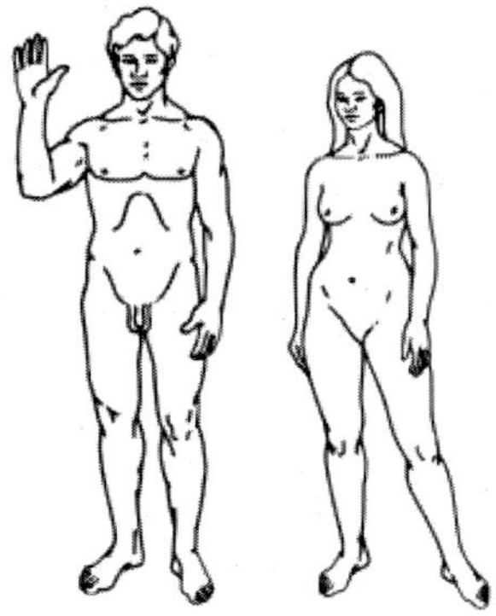 *Homo sapiens sapiens* (Pioneer plaque) Omo1, Omo2 (Ethiopia, Omo river) are the earliest fossil evidence for anatomically modern *Homo sapiens* [11].
160 ka	*Homo sapiens* (*Homo sapiens idaltu*) in Ethiopia, Awash River, Herto village, practice mortuary rituals and butcher hippos. Potential earliest evidence of anatomical and behavioral modernity consistent with the continuity hypothesis including use of red ochre and fishing.
150 ka	Mitochondrial Eve is a woman who lived in East Africa. She is the most recent female ancestor common to all mitochondrial lineages in humans alive today. Note that there is no evidence of any characteristic or genetic drift that significantly differentiated her from the contemporary social group she lived with at the time. Her ancestors were *Homo sapiens* as were her contemporaries.
90 ka	Appearance of mitochondrial haplogroup L2.
70 ka	Behavioral modernity according to the "great leap forward" theory [12].
60 ka	Appearance of mitochondrial haplogroups M and N, which participate in the migration out of Africa. Homo sapiens that leave Africa in this wave start interbreeding with the Neanderthals they encounter [13].
50 ka	Migration to South Asia. M168 mutation (carried by all non-African males). Beginning of the Upper Paleolithic. mt-haplogroups U, K.
40 ka	Migration to Australia [14] and Europe (Cro-Magnon).
25 ka	The independent Neanderthal lineage dies out. Y-Haplogroup R2; mt-haplogroups J, X.
12 ka	Beginning of the Mesolithic/Holocene. Y-Haplogroup R1a; mt-haplogroups V, T. Evolution of light skin in Europeans (SLC24A5). *Homo floresiensis* dies out, leaving *Homo sapiens* as the only living species of the genus *Homo*.

References

[1] Richard Dawkins (2004). *The Ancestor's Tale* page 136.

[2] Raauma R, Sternera K (2005) Catarrhine primate divergence dates estimated from complete mitochondrial genome. *J. Hum. Evol.* 48, 237-257.

[3] Perlman D (2001). Fossils From Ethiopia May Be Earliest Human Ancestor. *National Geographic News*.

[4] Bermúdez de Castro JM et al. (1997). A Hominid from the Lower Pleistocene of Atapuerca, Spain: Possible Ancestor to Neandertals and Modern Humans. *Science* 276, 1392-1395.

[5] Green RE, Krause J, Ptak SE et al. (2006). Analysis of one million base pairs of Neanderthal DNA. *Nature* 16, 330–336.

[6] Britten RJ (2002). Divergence between samples of chimpanzee and human DNA sequences is 5%, counting indels. *Proc. Natl. Acad. Ss. USA* 99, 13633–5.

[7] Pollard KS (2009). What makes us human? *Sc Am*. 300-5, 44–49.

[8] Krause J, Lalueza-Fox C, Orlando L et al. (2007). The derived FOXP2 variant of modern humans was shared with Neandertals. *Curr. Biol.*17, 1908–12.

[9] Mounier A, François M, Silvana C (2009). Is Homo heidelbergensis a distinct species? New insight on the Mauer mandible. *J. Hum. Evol.* 56, 219-246.

[10] Mendez F, Krahn T, Schrack B, et al. (2013). An African American paternal lineage adds an extremely ancient root to the human Y chromosome phylogenetic tree. *Am. J. Hum. Genet.* 92, 454.

[11] Hopkin M (2005). Ethiopia is top choice for cradle of *Homo sapiens*. *Nature News*.

[12] Diamond J (1992). *The Third Chimpanzee*. Harper Perennial. pp. 47–57.

[13] Rincon P (2010). Neanderthal genes' survive in us. *BBC News* (BBC).

[14] Bowler JM, Johnston H, Olley JM, Prescott JR, Roberts RG et al. (2003). New ages for human occupation and climatic change at Lake Mungo, Australia. *Nature* 421, 837–40.

APPENDIX III

GLOSSARY OF HUMAN EVOLUTION

WALKING TALL

- Humans are really just a peculiar African ape - we share about 98% of our DNA with chimpanzees, our closest living relatives.
- Genetics and fossil evidence hint that we last shared a common ancestor 7 to 10 million years ago - even if we continued hybridising long after.
- At around 6 million years ago, the first apes to walk on two legs appear in the fossil records. Australopithecines and other early humans were no bigger than chimps and had similar-sized brains.
- Aside from our large brain, bipedalism is perhaps the most important difference between humans and apes, as it freed our hands to use tools.
- Bipedalism may have evolved when drier conditions shrank dense African forests. It must have allowed our ancestors to spot predators from further away, reach hanging fruit from the ground, and reduce exposure to sunlight.
- Evidence that Australopithecines walked upright includes analysis of the shape of their bones and fossilised footprints.
- One famous member of the species *Australopithecus afarensis* is the remarkably complete fossil found by palaeaoanthropologist Donald Johanson in Hadar, Ethiopia in 1974. The 3.2-million-year-old fossil

was named Lucy, after the Beatles' song *Lucy in the Sky with Diamonds*.

- She stood around 1.1 metres (3.5 feet) tall and although she walked on two legs, she probably had a less graceful gait than us, since she walked with them bent.
- Scientists have modelled her gait using computers. Their characteristic long arms and curved fingers suggest that at least some Australopithecines were still good climbers.
- Hundreds of other fossils of *Australopithecus afarensis* have now also been discovered. Other related early human species include *Australopithecus africanus* - such as the Taung child - 3.5-million-year-old *Kenyanthropus platyops,* 5.8-million to 4.4-million-year-old *Ardipithecus*, 5.8-million-year-old*Orrorin tugenensis* and 6 million year old *Sahelanthropus tchadensis.*

TOOL UP

- Australopithecines are thought to be the ancestors of *Homo*, the group to which our own species, *Homo sapiens,* belongs.
- Australopithecines may also have given rise to another branch of hominid evolution - the vegetarian *Paranthropus* species.
- Around 2.7 million years ago, species such as *Paranthropus bosei* in east Africa evolved to take advantage of the dry grasslands. This included the development of enormous jaws and chewing muscles for grinding up tough roots and tubers.
- By 2.4 million years ago, *Homo habilis* had appeared - the first recognisably human-like hominid to appear in the fossil record - which lived alongside *P. bosei.*
- Their bodies were around two-thirds the size of ours, but their brains were significantly larger than Australopithecines with a volume of about 600 cubic centimetres.
- *H. habilis* had much smaller teeth and jaws than *Paranthropus* and was probably the first human to eat large quantities of meat.
- This meaty diet, acquired through scavenging, may have provided energy required to kick-start an increasing brain size.
- A mutation that weakened jaw muscles and gave our brains more space to grow may also lie behind the big brains we have today.

- *H. habilis* - which means "handy man" - was also the first early human to habitually create tools and use them to break bones and extract marrow.
- This tool-making tradition, known as Oldowan, lasted virtually unchanged for a million years.
- Oldowan tools were made by breaking an angular rock with a "hammerstone" to give simple, sharp-edged stone flakes for chopping and slicing.
- Despite their own increases in brain size, the *Paranthropus* group of species had become extinct by 1.2 million years ago.
- Some experts speculate that it was learning to work as a team against predators that gave *Homo* the edge.

Modern Looker

- At around 1.65 million years ago, another early human, *Homo ergaster*, started to create tools in a slightly different fashion. This so-called Acheulean tradition was the tool-making technology used for nearly the entire Stone Age, and practiced until 100,000 years ago. Acheulean tools, such as hand axes and cleavers, were larger and more sophisticated than their predecessors'. They may have been status symbols as well as tools.
- *Homo ergaster* first appeared in Africa around 2 million years ago, and in many ways resembled us. Though they had brow ridges, they had lost the stoop and long arms of their ancestors.
- They may have been even more slender than us and were probably well-adapted to running long distances.
- Some experts believe that they were the first to sport largely hairless bodies, and to sweat, though another theory puts our hairlessness down to an aquatic phase.
- One famous example of a more modern looking early human is the Turkana boy, a teenager when he died, 1.6 million years ago in Kenya. The shape of this fossil showed that the human pelvis had reached today's narrow proportions.
- Combined with the growing size of the human head and brain, this had far-reaching implications: human women now need help for a

successful birth; and human babies are born earlier, and need a longer period of childhood care, than those of apes.

- Meat-eating, however, may have allowed us to become early weaners.
- *H.ergaster* may have been the first early human to leave Africa. Bones dated to around 1.75 million years ago have been found in Dmanisi in Georgia.
- Shortly afterwards, *Homo erectus* appeared - the first early human whose fossils have been seen in large numbers outside of Africa.
- The first specimen discovered, a single cranium, was unearthed in Indonesia in 1891.
- *H.erectus* was highly successful, spreading to much of Asia between 1.8 and 1.5 million years ago, and surviving as recently as 27,000 years ago.
- *H.erectus*, with a brain volume of around 1000 cm^3 would have interacted with modern humans.
- They may have been the first people to take to the seas and habitually hunt prey such as mammoths and wild horses, although there is some debate about this. They may also have harnessed the use of fire and built the first shelters.
- In 2004, the remains of a tiny and mysterious human species, that may have lived as recently as 13,000 years ago, was discovered on an Indonesian island.
- More bones of the "hobbit", or *Homo floresiensis*, were uncovered in 2005.
- Some studies suggest it had an advanced brain and was unequivocally a separate species - but others argue that these people were modern humans suffering from a genetic disorder.

FIRST EUROPEANS

- Early human fossil evidence from Spain, dating to around 780,000 years ago, points to the first known Europeans.
- Stone tools have also been found in England from around 700,000 years ago, attributed to *Homo antecessor* or *Homo heidelbergensis.*
- More recently, 325,000-year-old *H. heidelbergensis* tracks were discovered preserved on an Italian volcano.

- Some of the biggest collections of hominid remains ever found are from Boxgrove in England and Atapuerca in Spain.
- Experts believe that these humans may have had ears equipped to detect nuances of human speech, whether or not they had simple language.
- Some palaeoanthropologists believe that *H. heidelbergensis* evolved into our own species in Africa, whilst in Europe, the Neanderthals emerged as a separate species.
- The Neanderthals were found across Europe, between 200,000 and 28,000 years ago. Though they still possessed pronounced brow ridges and were more thick-set, these people largely resembled us.
- Neanderthals were nimble-fingered, and matured at a similar age to us. Their brains were even slightly larger.
- It is not known if the Neanderthals had developed simple language. But they did possess some aspects of our culture, such as ritual burying of the dead; creating art; using tools to attack each other; and complex hunting methods - as evidenced by a remarkable butchery site in the UK.
- Experts disagree about whether the Neanderthals hybridised with humans or not, or if our arrival killed them. Plunging temperatures, free trade and poor memory may all have contributed towards their extinction.

OUT OF AFRICA

- There are several competing theories about how all these early humans are related to us today.
- Most widely accepted is the "Out of Africa" hypothesis. This holds that ancient humans evolved exclusively in Africa, then spread across the world in two migration waves.
- The migration of *H. erectus* across Eurasia made up the first wave. Later, our own species evolved in Africa and fanned out in a second wave 200,000 years ago. These new people totally replaced *H. erectus* in Asia and the Neanderthals in Europe.
- Advocates of the multiregional hypothesis instead believe that early humans started to leave Africa around 2 million years ago, and were never totally replaced by recent migrants. They believe these far-flung

hominids exchanged genes and interbred, slowly evolving into modern humans - in many places, simultaneously.

- Through gene flow, modern characteristics such as large brains gradually spread, it is suggested. Some fossils seem to support the multiregional hypothesis. *H. erectus* skulls in Asia, for example, have similarly flat cheek and nasal regions as people there today do.
- Most - but not all - genetic evidence appears to back the Out of Africa hypothesis.
- There is surprisingly little variation in the mitochondrial DNA (mDNA) of different people today, which suggest that humans evolved recently from a small ancestral population. In addition, the variation of mDNA in Africans is greater than elsewhere, suggesting that people have been evolving there for longer.
- We may all be descended from a single African woman - dubbedMitochondrial Eve - within the last 200,000 years. Male Y-chromosome DNAhints at a single male progenitor, too. Fewer than 50 people could have given rise to the entire population of Europe, experts believe.

CULTURAL REVOLUTION

- The earliest anatomically modern humans are though to have arrived around 200,000 years ago.
- These fossils show a rounded braincase and flatter face. Their brains had reached modern proportions of about 1350 cm^3.
- Two skulls found in Ethiopia make up the oldest modern human remains known, at 195,000 years old.
- Modern humans had made it to Asia by 90,000 years ago, Australia by 60,000 years ago, Europe and the Arctic by 40,000 years ago, and the Americas by 12,000 years ago.
- Throughout history, tool use appears to have progressed slowly - once innovations were made, they lasted millions of years barely altering.
- Around 50,000 years ago something changed, and culture started to develop at a much more rapid rate.
- Modern humans habitually began innovating new tools types, burying their dead, creating jewellery, developing sophisticated hunting

techniques such as pitfall traps, using animal skins for clothing, decorating their bodies, and creating art and cave paintings.

- Although some of these traits appeared earlier, they seem to have only have been used sporadically until this time.
- These changes may have been linked to increasing brain size or the way we thought - or could also be due to free trade, and the evolution of language and communication.
- The dawn of human civilisation has been dated to around 30,000 years ago. The earliest agriculture and domestication of species is known only as recently as 10,000 years ago. The first human cities appeared in Mesopotamia around 4,000 years ago.
- Are we still evolving today? If so, how will we evolve in the future? Some argue that humans have evolved little in the last 50,000 years - but other studies suggests that thousands of genes have changed since then.
- We may even be on the verge of the next step of human evolution - the human global "superorganism."

INDEX

A

B

C

D

E

F

G

H

I

J

N

O

P

T

U

V

W

Y